Microbiology

DeMYSTiFieD®

DeMYSTiFieD® Series

Accounting Demystified
Advanced Calculus Demystified
Advanced Physics Demystified
Advanced Statistics Demystified
Algebra Demystified
Alternative Energy Demystified
Anatomy Demystified
Astronomy Demystified
Audio Demystified
Biology Demystified
Biotechnology Demystified
Business Calculus Demystified
Business Math Demystified
Business Statistics Demystified
C++ Demystified
Calculus Demystified
Chemistry Demystified
Chinese Demystified
Circuit Analysis Demystified
College Algebra Demystified
Corporate Finance Demystified
Databases Demystified
Data Structures Demystified
Differential Equations Demystified
Digital Electronics Demystified
Earth Science Demystified
Electricity Demystified
Electronics Demystified
Engineering Statistics Demystified
Environmental Science Demystified
Ethics Demystified
Everyday Math Demystified
Fertility Demystified
Financial Planning Demystified
Forensics Demystified
French Demystified
Genetics Demystified
Geometry Demystified
German Demystified
Home Networking Demystified
Investing Demystified
Italian Demystified
Japanese Demystified
Java Demystified

JavaScript Demystified
Lean Six Sigma Demystified
Linear Algebra Demystified
Logic Demystified
Macroeconomics Demystified
Management Accounting Demystified
Math Proofs Demystified
Math Word Problems Demystified
MATLAB® Demystified
Medical Billing and Coding Demystified
Medical Terminology Demystified
Meteorology Demystified
Microbiology Demystified
Microeconomics Demystified
Nanotechnology Demystified
Nurse Management Demystified
OOP Demystified
Options Demystified
Organic Chemistry Demystified
Personal Computing Demystified
Pharmacology Demystified
Philosophy Demystified
Physics Demystified
Physiology Demystified
Pre-Algebra Demystified
Precalculus Demystified
Probability Demystified
Project Management Demystified
Psychology Demystified
Quality Management Demystified
Quantum Mechanics Demystified
Real Estate Math Demystified
Relativity Demystified
Robotics Demystified
Sales Management Demystified
Signals and Systems Demystified
Six Sigma Demystified
Spanish Demystified
sql Demystified
Statics and Dynamics Demystified
Statistics Demystified
Technical Analysis Demystified
Technical Math Demystified
Trigonometry Demystified

Microbiology
DeMYSTiFieD®

Tom Betsy, D.C.

Jim Keogh, R.N.

Second Edition

New York   Chicago   San Francisco   Lisbon   London   Madrid   Mexico City
Milan   New Delhi   San Juan   Seoul   Singapore   Sydney   Toronto

The McGraw·Hill Companies

Cataloging-in-Publication Data is on file with the Library of Congress.

McGraw-Hill books are available at special quantity discounts to use as premiums and sales promotions, or for use in corporate training programs. To contact a representative please e-mail us at bulksales@mcgraw-hill.com.

Microbiology DeMYSTiFieD®, Second Edition

2 3 4 5 6 7 8 9 0   DOC/DOC   1 9 8 7 6 5 4

ISBN 978-0-07-176109-3
MHID 0-07-176109-8

Sponsoring Editor Judy Bass	**Copy Editor** James Madru	**Art Director, Cover** Jeff Weeks
Acquisitions Coordinator Bridget Thoreson	**Proofreader** Laura Bowman	**Cover Illustration** Lance Lekander
Editing Supervisor David E. Fogarty	**Production Supervisor** Richard C. Ruzycka	
Project Manager Nidhi Chopra, Cenveo Publisher Services	**Composition** Cenveo Publisher Services	

This book is dedicated to Shelley, Juliana, and Thomas.
Thank you for your laughter and smiles and for tolerating the nights and weekends
Dad was away working on the manuscript.
To my students whose love of education inspires me and makes teaching the best job in the world.

Tom Betsy

This book is dedicated to Anne, Sandy, Joanne, Amber-Leigh Christine, Shawn, Eric, and Amy,
without whose help and support this book couldn't have been written.

Jim Keogh

About the Authors

Dr. Tom Betsy is a professor at Bergen Community College and a chiropractor in New Jersey. He has also written *Schaum's Outline of Pathophysiology*. He is widely respected for his unique way of simplifying complex topics.

Jim Keogh is a registered nurse and has written books in McGraw Hill's DeMYSTiFieD Series and the Schaum's Outline Series. These include: *Pharmacology DeMYSTiFieD, Microbiology DeMYSTiFieD, Medical-Surgical Nursing DeMYSTiFieD, Medical Billing and Coding DeMYSTiFieD, Schaum's Outline of ECG Interpretations, Schaum's Outline of Medical Terminology*, and *Schaum's Outline of Emergency Nursing*. Jim Keogh, RN, A.A.S., BSN, MBA is a former member of the faculty at Columbia University and is a member of the faculty of New York University.

Contents

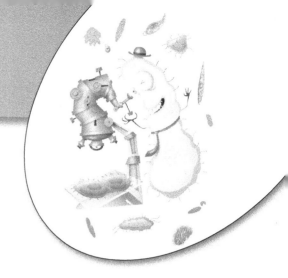

Acknowledgments

To my parents Walter and Patricia, thank you for always being there. To Joan and Ronnie Sisto; Joan has assisted me for most of my professional career. I thank you both very much.

Tom Betsy

Introduction

You hear the words *germ, bacteria,* and *virus* and you might cringe, then run for the nearest sink to wash your hands because these words bring back memories of when you caught a cold or the flu, which was not a pleasant experience. Germs, bacteria, viruses, and other microscopic organisms are called microorganisms, or microbes for short. And as you'll learn throughout this book, some microbes bring about disease while others help fight disease.

Think for a moment. Right now there are thousands of tiny microbes living on the tip of your finger in a world that is so small that it can only be visited by using a microscope. In this book we'll show you how to visit this world and how to interact with these tiny creatures that call the tip of your finger home.

The microscopic world was first visited in the late 1600s by Dutch merchant and amateur scientist Antoni van Leeuwenhoek. He was able to see living microorganisms by using a single-lens microscope. We've come a long way since van Leeuwenhoek's first visit. Today scientists are able to see through some microbes and study the organelles that bring them to life.

It wasn't until the Golden Age of Microbiology between 1857 and 1914 when scientists, such as the likes of Louis Pasteur and Robert Koch, made a series of discoveries that rocked the scientific community. During this period scientists identified microbes that caused diseases and learned how to cure those diseases and then prevent them from occurring through the use of immunization.

Scientists were able to achieve these remarkable discoveries by using culturing techniques to grow colonies of microbes in the laboratory. Once microbes could be grown at will, scientists focused their experiments on ways to slow that growth and stop microbes in their tracks—killing the microbe and curing the disease caused by the microbe.

Culturing microbes is central to the study of microbiology, and you'll be using many of the same culturing techniques to colonize microbes in your college laboratory. We provide step-by-step instructions on how to do this.

You'll find it difficult to live without the aid of microbes. For example, living inside your intestines are colonies of microorganisms. Just this thought is enough to make your skin crawl. As frightful as this thought might be, these microbes assist your body in digesting food. That is, you might have difficulty digesting some foods if you killed these microbes.

Microbes in your intestines are beneficial to you as long as they remain in your intestines. However, you'll become very ill should they decide to wander into other parts of your body. Don't become too concerned because these microbes tend to stay at home unless your intestines are ruptured as a result of trauma, which is when they like to go exploring.

By the end of this book you'll learn about the different types of microbes, how to identify them by using a microscope, and how to cultivate colonies of microbes.

A Look Inside

Following the step-by-step approach that is used in *Microbiology DeMYSTiFieD* helps you meet the challenges of learning microbiology. You will learn in a logical progression where topics are presented in an order in which many students like to learn them—starting with basic components and then gradually moving on to those that are a little more complex to understand.

Each chapter follows a time-tested formula that explains the topic in an easy-to-read style. You can then compare your knowledge with what you're expected to know by taking chapter tests and the final exam. There is little room for you to go adrift.

Chapter 1: The World of the Microorganism

You'll begin your venture into the microscopic world of microbes by learning the fundamentals. These are the terms and concepts that every student needs to understand before he or she can embark on more advanced topics, such as cultivating your own microbes.

In this chapter, you are introduced to the science of microbiology with a look back in time to a period when little was known about microbes except that people were dying. You'll also learn about the critical accomplishments that were made in microbiology that enable scientists to understand and develop cures for disease.

Chapter 2: The Chemical Elements of Microorganisms

Chemistry is a major factor in microbiology because microbes are made up of chemical elements. Scientists are able to destroy microbes by breaking down microbes into their chemical elements and then disposing of those elements.

Before you can understand how this process works, you must be familiar with the chemical principles that are related to microbiology. You'll learn about those chemical principles in this chapter.

Chapter 3: Observing Microorganisms

"Wash the germs from your hands!" That is the cry of every mom who knows that hand washing is the best way to prevent sickness. Most kids balk at hand washing simply because they can't see the germs on their hands.

We'll show you how to see germs and other microbes in this chapter by using a microscope. You'll learn everything you need to know to bring microbes into clear focus so you can see what you can't see with the naked eye.

Chapter 4: Prokaryotic Cells and Eukaryotic Cells

It is time to get up close and personal with two common microbe cells. These are prokaryotic cells and eukaryotic cells. These names are probably unfamiliar to you but they won't be by the time you're finished reading this chapter.

Prokaryotic cells are bacteria cells, and eukaryotic cells are cells of animals, plants, algae, fungi, and protozoa. Each carry out the six life processes required of every living thing. In this chapter you'll learn about how each carry out these life processes.

Chapter 5: Chemical Metabolism

"It is my slow metabolism! That's why I can't shed a few pounds." It is a great excuse for being unable to lose weight, but the reason our metabolism is slow is because we tend not to exercise enough.

In this chapter, you'll learn the biochemical reactions that change food into energy—called metabolism—and how the cell is able to convert nutrients into energy.

Chapter 6: Microbial Growth and Controlling Microbial Growth

You and microbes need nutrients, such as carbon, hydrogen, nitrogen, and oxygen, to grow. However, not all microbes need the same chemical nutrients.

For example, some require oxygen while others can thrive in an oxygen-free environment.

You'll learn in this chapter how to classify microbes by the chemical nutrients they need to survive. You'll also learn how to use this knowledge to grow and control the growth of microbes in the laboratory.

Chapter 7: Microbial Genetics

Just like us, microbes inherit genetic traits from previous generations of microbes. Genetic traits are instructions on how to enable everything to stay alive. Some instructions are passed along while other instructions aren't handed over to the next generation.

In this chapter, you'll learn how microorganisms inherit genetic traits from previous generations of microorganisms. Some of those traits show them how to identify and process food, how to excrete waste products, and how to reproduce.

Chapter 8: Recombinant DNA Technology

Who we are and what we are going to be are programmed into our genes. The same is true about microbes. This genetic information is encoded into DNA by the linking of nucleic acids in a specific sequence.

Genetic information can be reordered in a process called genetic engineering. You'll learn about genetic engineering and recombinant DNA using DNA technology in this chapter.

Chapter 9: Classification of Microorganisms

There are thousands of microbes and no two are identical, but many have similar characteristics. Microbiologists have spent years carefully observing microbes and organizing them into groups by their similarities.

You'll learn how microbes are classified in this chapter, which enables you to efficiently identify microbes that you see under a microscope.

Chapter 10: The Prokaryotes: Bacteria and Archaea

Bacteria are one of the most common microbes that you encounter. Some bacteria cause disease and other bacteria help you live by aiding in digestion. You can imagine there are many different kinds of bacteria, however all can be grouped into four divisions based on their cell walls.

Each division is divided into sections based on other characteristics such as oxygen requirements, motility, shape, and Gram stain reaction. In this chapter, you'll learn how to use these divisions and sections to identify bacteria.

Chapter 11: The Eukaryotes: Fungi, Algae, Protozoa, and Helminths

In this chapter you'll take a close look at kingdoms of Fungi, Protista, and Animalia. No, this hasn't anything to do with a mythical society. These are microbes that are commonly known as fungi, algae, protozoa, and helminths.

Eukaryotes are a type of microbes that are different from bacteria and viruses. However, they too are beneficial to us by supplying food, removing waste, and curing disease by being used as an antibiotic. And as bacteria, some eukaryotes also cause disease.

Chapter 12: Viruses, Viroids, and Prions

Probably one of the most feared microbes is a virus because many times there is little or nothing that can be done to kill it. Once you're infected, you can treat the symptoms, such as a runny nose and watery eyes, but otherwise you must let the virus run its course.

Did you ever wonder why this is the case? If so, then read this chapter for the answer and learn what a virus is, how viruses live, and the diseases they cause.

Chapter 13: Epidemiology and Disease

It's flu season and you can only hope that you don't become infected, otherwise you'll have ten days of chills, sneezing, and isolation. No one wants to come close to you for fear of catching the flu.

In this chapter, you'll learn about diseases like the flu and how diseases are spread. You'll also learn how to take simple precautions to control and prevent the spread of diseases.

Chapter 14: Immunity

Inside your body there is a war going on. An army of B cells, T cells, natural killer cells, and other parts of your immune system are on the defense seeking microbes to rip apart before any microbes can give you a runny nose, cough, and that feverish feeling.

The immune system is your body's defense mechanism that surrounds, neutralizes, and destroys foreign invaders before they can do harm. You'll learn about your immune system and how it gives you daily protection against invading microbes in this chapter.

Chapter 15: Vaccines and Diagnosing Diseases

Think about this: Each year millions of people pay their doctor to inject them with the flu virus. On the surface it doesn't make sense, but after reading this chapter you'll find that it makes perfect sense because the injection is a vaccination against the flu.

Vaccines prevent you from catching certain diseases because the vaccine has elements of the disease that trigger your body to create antibodies to the disease. You'll learn about vaccines and antibodies in this chapter.

Chapter 16: Antimicrobial Drugs

"Doc, give me a pill to knock out whatever is causing me to be sick!" Most of us would like to say this whenever we come down with an illness. All we want is the magic pill that makes us feel better. Sometimes that magic pill—or injection—contains a microbe that seeks out and destroys pathogenic microbes, which are disease-causing microbes.

In this chapter you'll learn about antimicrobial drugs that are given as chemotherapy to cure disease. These antimicrobial drugs contain microbes that kill other microbes.

Microbiology

DeMYSTiFieD®

chapter 1

The World of the Microorganism

In this chapter, you'll be introduced to fundamental concepts of microbiology and microorganisms and characteristics of the different species of microorganisms.

CHAPTER OBJECTIVES

In this chapter, you will

- Learn about the types of microorganisms
- Become familiar with the naming and classification of microorganisms
- Examine the immune system
- Begin to understand how a microscope is used

An *organism* is a living thing that requires food to sustain life. Food is ingested and broken down by the organism into energy and nutrients required for the organism to function. Undigested food is excreted as waste. An organism reproduces to propagate the species.

A *microorganism* is a very small organism that lives outside and inside larger organisms such as the human body. A microorganism cannot be seen without the assistance of magnifying devices such as a microscope because of the microorganism's tiny size. However, the effects of a microorganism on the human body can be felt through signs or symptoms of infection. An infection is the human body's adverse response to the presence of a pathogenic microorganism inside the body. For example, an infection caused by a microorganism called a *virus* may cause watery eyes and increase excretion of mucus from the nose. This disorder is commonly referred to as a *head cold*. Increased fluid production by the body is the way the immune system retaliates against the virus. The goal is to flush the virus from the body. Other microorganisms live in our body without doing us any harm. These microbes are considered "normal flora" and are normally found in the upper respiratory tract, gastrointestinal tract, and vagina.

Microbiology is the study of microorganisms. By identifying microorganisms and how they function, scientists are able to develop interventions that prevent microorganisms from invading the body as well as medications that kill microorganisms inside the body, thus eliminating their adverse side effects on the body.

Types of Microorganisms

Unfriendly Microorganisms

An infection is caused by the infiltration of a disease-causing microorganism known as a *pathogen*. Some pathogenic microorganisms infect humans but not other animals and plants. Some pathogenic microorganisms that infect animals or plants also infect humans.

Pathogenic microorganisms make headlines and play an important role in history. Legendary gunfighter John "Doc" Holliday is famous for his escapades in the Wild West. He dodged countless bullets, showing that he was the best of the best when it came to gun fighting, yet *Mycobacterium tuberculosis* took down Doc Holliday quietly, without firing a shot. *M. tuberculosis* is the bacterium that causes tuberculosis (Fig. 1-1). This bacterium affects the lung tissue

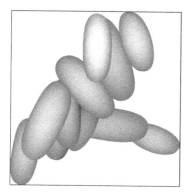

FIGURE 1-1 • *M. tuberculosis* is the bacterium that causes tuberculosis.

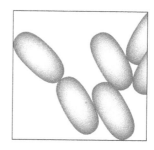

FIGURE 1-2 • *Y. pestis* is the microorganism that caused the black plague.

when droplets of respiratory secretions or particles of dry sputum from a person who is infected with the disease are inhaled by an uninfected person.

Yersinia pestis nearly conquered Europe in the fourteenth century with the help of the flea. *Y. pestis* is the microorganism that caused the black plague (Fig. 1-2) and killed more than 25 million Europeans. You might say that *Y. pestis* launched a sneak attack. First, it infected fleas that were carried into populated areas on the backs of rats. Rodents traveled on ships and then over land in search of food. Fleas jumped from rodents and bit people, transmitting the *Y. pestis* microorganism into the person's bloodstream.

In an effort to prevent the spread of *Y. pestis*, sailors entering Sicily's seaports had to wait 40 days before leaving the ship. This gave time for sailors to exhibit the symptoms of the black plague if the *Y. pestis* microorganism had infected them. Sicilians called this *quarantenaria*. Today we know it as *quarantine*. Sailors who did not exhibit symptoms were not infected and were free to disembark.

Campers and travelers sometimes become acquainted with *Giardia lamblia*, *Escherichia coli*, or *Entameba histolytica*. Travelers who become infected typically do not die but come down with a bad case of diarrhea.

Friendly Microorganisms

Not all microorganisms are pathogens. In fact, many microorganisms help to maintain homeostasis in our bodies and are used in the production of food and other commercial products. For example, *flora* are microorganisms found in our intestines that assist in the digestion of food and play a critical role in the formation of vitamins such as vitamins B and K. They help by breaking down large molecules into smaller ones.

What Is a Microorganism?

Microorganisms are the subject of *microbiology*, which is the branch of science that studies microorganisms. A microorganism can be one cell or a cluster of cells that can be seen only by using a microscope.

Microorganisms are organized into six fields of study: bacteriology, virology, mycology, phycology, protozoology, and parasitology.

Bacteriology

Bacteriology is the study of bacteria. *Bacteria* are prokaryotic organisms. A *prokaryote* is a one-celled organism that does not have a true nucleus. Many bacteria absorb nutrients from their environment, and some make their own nutrients by photosynthesis or other synthetic processes. Some bacteria can move freely in their environment with the help of flagella whereas others are stationary. Bacteria occupy space on land and can live in aquatic environments and in decaying matter. They can even cause disease. *Bacillus anthracis* is a good example. It is the bacterium that causes anthrax.

Virology

Virology is the study of viruses. A *virus* is an infectious agent composed of a nucleic acid core surrounded by a protein coat. Viruses are not considered living organisms. They lack independent metabolism and can only reproduce within the cells of a living host. An example of a virus is the *varicella-zoster virus* (Fig. 1-3), which is the virus that causes chickenpox in humans.

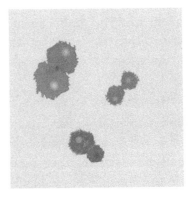

FIGURE 1-3 · The varicella-zoster virus causes chickenpox.

Mycology

Mycology is the study of fungi. A *fungus* is a eukaryotic organism, often microscopic, that absorbs nutrients from its external environment. Fungi are not photosynthetic but they do possess a cell wall for structure and support. A *eukaryote* is a microorganism whose cells have a nucleus, cytoplasm, and organelles. These include yeasts and some molds. *Tinea pedis*, better known as athlete's foot, is caused by the *Trichophyton rubrum* fungus.

Phycology

Phycology is the study of algae. *Algae* are eukaryotic photosynthetic organisms that transform sunlight into nutrients using photosynthesis. A *photosynthetic eukaryote* is a microorganism whose cells have a nucleus, nuclear envelope, cytoplasm, and organelles, called chloroplast, that are able to carry out photosynthesis.

Protozoology

Protozoology is the study of *protozoa*, animal-like single-cell microorganisms. Many protozoa obtain their food by engulfing or ingesting smaller organisms. An example is *Amoeba proteus*.

Parasitology

Parasitology is the study of parasites. A *parasite* is an organism that lives at the expense of another organism or host. Parasites that cause disease are called *pathogens*. Examples of parasites are bacteria, viruses, protozoa, and many animals such as round worms, flatworms, and arthropods (insects).

Still Struggling

Keeping track of the study of organisms can be difficult. Remember: *Bacteriology* is the study of bacteria, and *bacteria* are prokaryotic organisms. *Virology* is the study of viruses, and *viruses* are infectious entities composed of a nucleic acid surrounded by a protein coat. *Mycology* is the study of fungi, and *fungi* are eukaryotic organisms that absorb nutrients from their external environment. *Phycology* is the study of algae, and *algae* are eukaryotic photosynthetic organisms. *Protozoology* is the study of protozoa, and *protozoa* are motile (usually) single-celled eukaryotic organisms. *Parasitology* is the study of parasites, and a *parasite* is an organism that lives and benefits at the expense of another host organism.

What's in a Name: Naming and Classifying

Carl Linnaeus developed the system for naming organisms in 1735. This system is referred to as *binominal nomenclature*. Each organism is assigned two latinized names because Latin or Greek was the traditional language used by scholars. The first name is called the *genus*. The second name is called the *specific epithet*, which is the name of the *species* and is not capitalized. The genus and the epithet appear underlined or italicized.

The name itself describes the organism. For example, *Staphylococcus aureus* is a very common bacterium. *Staphylococcus* is the genus, and *aureus* is the epithet (or species). In this case, the genus describes the appearance of the cells. *Staphylo* means a clustered arrangement of the cells, and *coccus* signifies that the cells are spheres. In other words, this means a cluster of sphere-like cells. *Aureus* is the Latin word for golden, which means that the cluster of sphere-like cells has a golden hue.

Sometimes an organism is named for a researcher, as is the case with *Escherichia coli* (Fig. 1-4), better known as *E. coli*. The genus is *Escherichia*, which is named for Theodor Escherich, a leading microbiologist. The epithet is *coli*, which implies that the bacterium lives in the colon (large intestine).

Organisms were classified into either the animal kingdom or the plant kingdom before the scientific community discovered microorganisms in the seventeenth century. It was at that time scientists realized that this classification system was no longer valid.

Carl Woese developed a new classification system that arranged organisms according to genetic relationships. However, it wasn't until 1978 that scientists

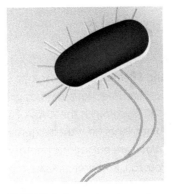

FIGURE 1-4 • *E. coli* is a bacterium that lives in the colon.

agreed on the new system for classifying organisms, and it took 12 more years before the new system was published. The new system utilizes domains which are longer than kingdoms.

Woese divided three classification groups into three *domains*:

- *Eubacteria*—Bacteria that have peptidoglycan in their cell walls. (Peptidoglycan is a large meshlike molecule composed of repeating subunits of the sugar derivatives of *N*-acetylglucosamine (NAG) and *N*-acetylmuramic acid (NAM), as well as several different amino acids.
- *Archaea*—Prokaryotes that have cell walls are not composed of peptidoglycan.
- *Eukarya*—Organisms from the following kingdoms:
  - *Protista*—Algae, protozoa, slime molds (*Note*: This is in the process of changing.)
  - *Fungi*—One-celled yeasts, multicellular molds, and mushrooms
  - *Plantae*—Moss, conifers, ferns, flowering plants
  - *Animalia*—Insects, worms, sponges, and vertebrates

How Small Is a Microorganism?

Microorganisms are measured using the metric system, which is shown in Table 1-1. In order to give you some idea of the size of a microorganism, let's compare a microorganism with things that are familiar to you:

German shepherd	1 meter
Human gamete (egg) from a female ovary	1 millimeter
A human red blood cell	100 micrometers

TABLE 1-1 Quantity and Length: Metric and English Equivalents

Unit	Fraction of Standard	English Equivalent
Meter (m)		3.28 feet
Centimeter (cm)	0.01 m = 10^{-2}	0.39 inch
Millimeter (mm)	0.001 m = 10^{-3}	0.039 inch
Micrometer (μm)	0.000001 m = 10^{-6}	0.000039 inch
Nanometer (nm)	0.000000001 m = 10^{-9}	0.000000039 inch

A typical bacterium cell	10 micrometers
A virus	10 nanometers
An atom	0.1 nanometer

Your Body Fights Back

Immunology is the study of how an organism defends itself against invading foreign materials, such as pathogenic microorganisms.

An example of how our bodies can protect us is *phagocytosis*. Phagocytosis is the ability of a cell to engulf and digest solid materials by the use of *pseudopods*, or "false feet." Phagocytosis was discovered in 1880 by Russian zoologist Elie Metchnikoff, who was one of the first scientists to study immunology. Metchnikoff studied the body's defense against disease-causing agents and invading microorganisms. He discovered that leukocytes (white blood cells) defend the body by engulfing and eating the invading microorganism.

Drugs: Send in the Cavalry

Invading microorganisms activate your body's immune system and result in various signs and symptoms. In an effort to help your immune system fight infection, physicians prescribe drugs called *antibiotics* that contain one or more antimicrobial agents that specifically combat bacteria. Other *antimicrobial agents* include antiviral medicines, antifungal agents, and antiparasitic medicines that combat invading pathogens.

One of the most commonly used antimicrobial agents is *penicillin*. Penicillin is made from *Penicillium*, which is a mold that secretes materials that interfere with the synthesis of the cell walls of bacteria. This causes *lysis*, or destruction of the cell wall, and kills the invading microorganism.

Immunity: Preventing a Microorganism Attack

Our bodies have a wide range of responses in the fight against pathogens. These responses are of two types: nonspecific and specific resistance. *Resistance* is the ability of the body to ward off disease. The lack of resistance is called *susceptibility*. When your immune system is compromised, you become susceptible to pathogens invading your body, where they multiply and subsequently make you sick.

Generally, your first line of defense is to use mechanical and chemical means to prevent a pathogen from entering your body. This initial defense is nonspecific.

Skin is the primary mechanical means to fight pathogens; it acts as a barrier between the pathogen and the internal structures of your body. Mucous membranes are another mechanical barrier; they help remove pathogens using tears and saliva. Mucus-producing cells and cilia in the upper respiratory track help keep pathogens out of the lungs. Urination, defecation, and vomiting are other mechanical means to combat pathogens by forcefully removing them from your body.

Other nonspecific responses include chemicals produced by the body that cause shifts in pH. Skin sebum is an example. Sebum is a thick substance secreted by the sebaceous glands; it consists of lipids and cellular debris that have a low pH, enabling it to chemically destroy a pathogen.

Sweat and saliva contain the enzyme lysozyme, which attacks the cell walls of bacteria. Hyaluronic acid found in areolar connective tissue sets up a chemical barrier that restricts a pathogen to a localized area of the body. Likewise, gastric juice and vaginal secretions have a low pH that creates a natural barrier to many kinds of pathogens.

History of the Microscope

Diseases are less baffling today than they were centuries ago, when scientists and physicians were clueless as to what caused disease. Imagine for a moment that a close relative became ill. One day she was well, and the next day she was sick for no apparent reason. Soon she was dead if her body couldn't fight the illness. You couldn't see whatever attacked her—and neither could the doctor.

Zacharias Janssen

In 1590, Zacharias Janssen developed the first compound microscope in Middleburg, Holland. Janssen's microscope consisted of three tubes. One tube served as the outer casing and contained the other two tubes. At either ends of the inner tubes were lenses used for magnification. Janssen's design enabled scientists to adjust the magnification by sliding the inner tubes. This enabled scientists to enlarge the image of a specimen three and nine times the specimen's actual size.

Robert Hooke

In 1665, Robert Hooke, an English scientist, popularized the use of the compound microscope when he placed the lenses over slices of cork and viewed little boxes that he called *cells*. It was his discovery that led to the development of cell theory in the nineteenth century by Mathias Schleiden, Theodor Schwann, and Rudolf Virchow. *Cell theory* states that all living things are composed of cells.

Antoni van Leeuwenhoek

Hooke's experiments with his crude microscope inspired Antoni van Leeuwenhoek to further explore the micro world. Van Leeuwenhoek, an amateur lens grinder, improved Hooke's microscope by grinding lenses to achieve magnification. His microscope required one lens. With his improvement, van Leeuwenhoek became the first person to view a living microorganism, which he called *Animalcules*.

This discovery took place during the 1600s, when scientists believed that organisms generated spontaneously and did not come from another organism. This sounds preposterous today, but back then scientists were just learning that a cell was the basic component of life.

How Do Organisms Appear?

Francesco Redi

Italian physician Francesco Redi developed an experiment that demonstrated that organisms do not appear spontaneously. He filled jars with rotting meat. Some jars he sealed, and others he left opened. Those that were open eventually contained maggots, which is the larval stage of the fly. The other jars did not contain maggots because flies could not enter the jars to lay their eggs on the rotting meat.

Redi's critics stated that air was the ingredient required for spontaneous generation of an organism; air was absent from the sealed jar and, consequently, no spontaneous generation could occur. (Fig. 1-5). Redi repeated the experiment, this time placing a screen over the opened jars. Since this prevented flies from entering the jar, there weren't any maggots on the rotting meat.

Prior to Redi's experiments, early scientists did not understand how to fight disease. Redi's discovery provided them with more information. They used Redi's findings to conclude that killing the microorganism that caused a disease

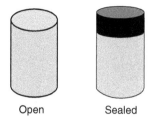

Open   Sealed

FIGURE 1-5 · No spontaneous generation occurred in
the sealed jar.

could prevent the disease from occurring. A new microorganism could be generated only by the reproduction of an existing microorganism. Kill the first
microorganism, and you will not have new microorganisms, the theory went—
you could stop the spread of the disease. Scientists called this the *theory of biogenesis*, which states that a living cell is generated from another living cell.

Louis Pasteur

Although the theory of biogenesis disproved spontaneous generation, spontaneous generation was hotly debated among the scientific community until 1861,
when Louis Pasteur, a French scientist, resolved the issue once and for all.
Pasteur showed that microorganisms were in the air. He proved that sterilized
solutions became contaminated once they were exposed to the air.

Pasteur came to this conclusion by boiling beef broth in several short-necked
flasks. Some flasks were left open to cool. Other flasks were sealed after boiling.
The opened flasks became contaminated with microorganisms, whereas no
microorganisms appeared in the closed flasks. Pasteur concluded that airborne
microorganisms had contaminated the opened flasks.

In a follow-up experiment, Pasteur placed beef broth in an open long-necked
flask. The neck was bent into an S shape. Again, he boiled the beef broth and
let it cool. The S-shaped neck trapped the airborne microorganisms (Fig. 1-6).
The beef broth remained uncontaminated even after months of being exposed
to the air. The very same flask containing the original beef broth exists today in
the Pasteur Institute in Paris and still shows no sign of contamination. Pasteur's
experiments validated that microorganisms are not generated spontaneously.

Based on Pasteur's findings, a concerted effort was launched to improve
sterilization techniques to prevent microorganisms from reproducing. *Pasteurization*, one of the best known sterilization techniques, was developed and
named for Pasteur. Pasteurization kills harmful microorganisms in milk, alcoholic

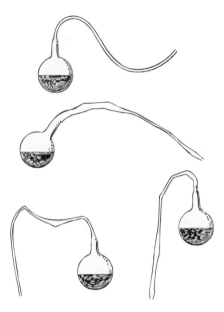

FIGURE 1-6 • Pasteur placed beef broth into a long-necked flask and then bent the neck into an S shape.

beverages, and other foods and drinks by heating them enough to kill most bacteria that cause spoilage.

John Tyndall and Ferdinand Cohn

The work of John Tyndall and Ferdinand Cohn in the late 1800s led to one of the most important discoveries in sterilization. They demonstrated that some microorganisms are resistant to certain sterilization techniques. Until their discovery, scientists had assumed that no microorganism could survive boiling water, which became a widely accepted method of sterilization. This was wrong. Some thermophiles, as well as endospores, resisted heat and could survive a bath in boiling water. This meant that there was not one magic bullet that killed all harmful microorganisms.

Germ Theory

Until the late 1700s, not much was really known about diseases except their impact. It seemed that anyone who came in contact with an infected person contracted the disease. A disease that is spread by exposure to the source of infection is called a *contagious disease*. The unknown agent that causes the disease is called a *contagion*. Today, we know that a contagion is a microorganism, but in the 1700s, many people found it hard to believe that something so small could cause such devastation.

Robert Koch

Opinions changed dramatically following Robert Koch's study of anthrax in the late 1800s. Koch observed that anyone who worked with or ingested animals that were infected with anthrax typically contracted the disease. In fact, people who simply inhaled the air around an infected animal were likely to inhale the anthrax bacterium spores and come down with the disease. Koch's investigations into anthrax led him to discover how microorganisms work.

Anthrax is caused by *Bacillus anthracis* (Fig. 1-7), which is a rod-shaped bacterium. *B. anthracis* can survive in a *dormant* or *inactive state* called an *endospore*. An endospore is not infectious. However, under the right conditions, the *B. anthracis* endospores can enter an active state, multiply rapidly, and become infectious. Endospores are structures produced by the genus *Bacillus* and *Clostridium* that are resistant to environmental stresses such as high heat, pH extremes, chemical exposure, and lack of water and nutrients.

The question that Koch raised is this: Would taking active *B. anthracis* from one animal and injecting it into a healthy animal cause the healthy animal to become ill with anthrax? If so, then he could prove that a microorganism actually was the cause of disease.

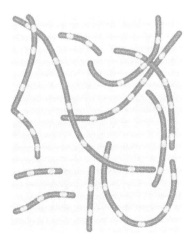

FIGURE 1-7 • *B. anthracis* multiply rapidly in the active state and become infectious.

B. anthracis was present in the blood of infected animals, so Koch removed a small amount of blood and injected it into a healthy animal. The animal came down with anthrax. He repeated the experiment by removing a small amount of blood from the newly infected animal and gave it to another healthy animal. It, too, came down with anthrax.

Koch expanded his experiment by cultivating *B. anthracis* on a slice of potato. He then exposed the potato to the right blend of air, nutrients, and temperature. Koch took a small sample of his homegrown *B. anthracis* and injected it into a healthy animal. The animal came down with anthrax.

Based on his findings, Koch developed the *germ theory*. The germ theory states that germs can enter and cause disease in other organisms. The disease-causing microorganism should be present in animals infected by the disease and not in healthy animals. The microorganism can be cultivated away from the animal and used to inoculate a healthy animal. The healthy animal then should come down with the disease. Samples of a microorganism taken from several infected animals are the same as the original microorganism from the first infected animal.

The four steps used by Koch to study microorganisms are referred to as *Koch's postulates*. Koch's postulates state the following:

1. The microorganism must be present in the diseased animal and not present in the healthy animal.
2. The suspected microorganism must be isolated and grown in a pure culture.
3. Symptoms of the disease should appear in the healthy animal after the healthy animal is inoculated with the culture of the microorganism.
4. The same microorganism must be isolated again from the diseased host. The new culture should be the same as the microorganism that was cultivated from the original diseased animal.

Koch's work with anthrax also developed techniques for growing cultures of bacteria. Because bacteria did not always grow well on potatoes, he tried growing them in a liquid medium that was hardened by gelatin. Although adding gelatin to media provided a better method for culture than using a potato, this method was also not ideal. Some bacteria could consume and digest the gelatin, and in addition, the gelatin would melt if heated above 28°C.

Around 1887, one of Koch's assistants, Walther Hesse, a bacteriologist and physician, suggested that they use agar instead of gelatin. This was the idea of his wife, Fannie Eilshemius. Fannie used agar, a polysaccharide from red

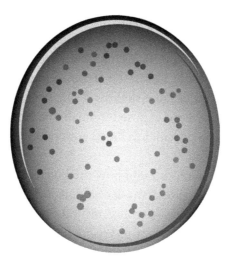

FIGURE 1-8 · A Petri dish is used to grow a culture of microorganisms.

algae, as a solidifying agent for her jellies. Agar was a better solidifying agent because most bacteria could not digest it, and it did not melt at temperatures under 100°C. Richard Petri improved on Koch's cultivating technique by placing the agar in a specially designed disk that was later to be called the *Petri dish* (Fig. 1-8). It, too, is still used today.

Koch did find that growth on gelatin inhibited the movement of microorganisms. As microorganisms reproduced, they remained together, forming a colony that made them visible without a microscope. The reproduction of microorganisms is called *colonizing.*

Vaccination

The variola virus which causes small pox was one of the most feared villains in the late 1700s. If variola didn't kill you, it caused pus-filled blisters that left deep scars that pitted nearly every part of your body. Cows also were susceptible to a variation of variola called *cowpox*. Milkmaids who tended to infected cows contracted cowpox and exhibited immunity to the smallpox virus.

Edward Jenner

Edward Jenner, an English physician, discovered something very interesting about both smallpox and cowpox in 1796. Those who survived smallpox never contracted smallpox again, even when they were later exposed to someone who was infected with smallpox. Milkmaids who contracted cowpox never caught smallpox, even though they were exposed to smallpox.

Jenner had an idea. He took scrapings from a cowpox blister found on a milkmaid, and using a needle, he scratched the scrapings into the arm of 8-year old James Phipps. Phipps became slightly ill when the scratch turned bumpy. Phipps recovered and then was exposed to smallpox. He did not contract smallpox because his immune system developed antibodies that could fight off both variola and vaccinia virus.

Jenner's experiment demonstrated how the body's own defense mechanism could be used to prevent disease by *inoculating* a healthy person with a tiny amount of the disease-causing microorganism. Jenner called this a *vaccination*, which is an extension of the Latin word *vacca* (cow). The person who received the vaccination became *immune* to the disease-causing microorganism.

Elie Metchnikoff

Elie Metchnikoff, a nineteenth-century Russian zoologist, was interested in Jenner's work with vaccinations. Metchnikoff explored the body's immune system to understand how it reacted to vaccinations. He discovered that white blood cells (*leukocytes*) engulf and digest microorganisms that invade the body. He called these cells *phagocytes*, which means "cell eating." Metchnikoff was one of the first scientists to study the new area of biology called *immunology*, the study of the immune system.

Killing the Microorganism

Great strides were made during the late 1800s in the development of antiseptic techniques. It began with a report by Hungarian physician, Ignaz Semmelweis, on a dramatic decline in childbirth fever (puerperal fever) when physicians used antiseptic techniques when delivering babies. Infections become preventable through the use of antiseptic techniques.

Joseph Lister

Joseph Lister, an English surgeon, developed one of the most notable antiseptic techniques. During surgery, he sprayed carbolic acid over the patient and then bandaged the patient's wound with carbolic acid–soaked bandages. Infection following surgery dropped dramatically when compared with surgery performed without spraying carbolic acid. *Carbolic acid*, also known as *phenol*, was one of the first surgical antiseptics.

Paul Ehrlich

Antiseptics prevented microorganisms from infecting a person, but scientists still needed a way to kill microorganisms after they infected the body. Scientists needed a "magic bullet" that cured diseases. At the turn of the nineteenth century, Paul Ehrlich, a German chemist, discovered the magic bullet. Ehrlich blended chemical elements into a concoction that, when inserted into an infected area, killed microorganisms without affecting the patient. Today we call Ehrlich's concoction a *drug*. Ehrlich's innovation has led to chemotherapy using synthetic drugs that are produced by chemical synthesis.

Alexander Fleming

Scientists from all around the world utilized Ehrlich's findings to develop drugs that could make infected patients well again. One of the most striking breakthroughs came in 1929, when Alexander Fleming discovered *Penicillin notatum*, the organism that synthesizes penicillin. *P. notatum* is a fungus that kills the *S. aureus* microorganism (Fig. 1-9) and similar microorganisms.

Fleming was growing cultures of *S. aureus*, a bacterium, in his laboratory, as well as conducting experiments with *P. notatim*, a mold. By accident, the *S. aureus* was contaminated with the *P. notatum*, causing the *Staphylococcus* to stop reproducing and die. *P. notatum* was the organism that produced the first true antibiotic called penicillin. An *antibiotic* is a substance that kills bacteria or inhibits the growth of microorganisms.

FIGURE 1-9 • *P. notatum* is a fungus that kills *S. aureus*.

TABLE 1-2	Scientists and Their Contributions	
Year	**Scientist**	**Contribution**
1590	Zacharias Janssen	Developed the first compound microscope.
1590	Robert Hooke	Observed nonliving plant tissue of a thin slice of cork.
1668	Francesco Redi	Discovered that microorganisms did not appear spontaneously. His contribution led to the finding that killing the microorganism that caused the disease could prevent the disease.
1673	Antoni van Leeuwenhoek	Invented the single-lens microscope. First person to view a living organism.
1798	Edward Jenner	Developed vaccinations against disease-causing microorganisms.
1887	Walther Hesse and Fannie Eilshemius	Used agar from red algae as a solidifying agent for culture media.

A summary of the scientists and their contributions can be found in Table 1-2.

The Method of Science

Microbiologists and other scientists utilize the scientific method to understand the natural world (Table 1-3). Scientists gather information based on observations and take an "educated guess" to explain these observations. This guess is

TABLE 1-3 The Scientific Method
• Observe a problem or have an idea.
• Develop a hypothesis.
• Set up experimentation to test the hypothesis.
• Gather results from the experiment.
• Analyze the information.
• If the data do not support the hypothesis, the hypothesis is rejected, and a new hypothesis is developed.
• If the data support the hypothesis, more rigorous tests are formulated and performed, and the hypothesis survives the testing.
• The hypothesis is accepted as a theory.

called a *hypothesis*. The next step in this process is to test the hypothesis. This is done by carefully gathering information from these observations through *experimentation*. After the information has been gathered, it is decided whether the information supports or does not support the hypothesis. If the hypothesis does not pass the test, it is rejected, and a new explanation or hypothesis is constructed. If the second hypothesis passes the test, more extreme testing is undertaken. This generalized approach is called the *deductive method*. The experimenter deduces a prediction from past accepted hypotheses. If the experiment is performed correctly, a certain result will occur. When using the *inductive method*, a generalized conclusion occurs after many examples are observed. Both types of methods are used by scientists to reach a conclusion. If the hypothesis survives the rigorous testing and experimentation of the scientific method, it is accepted as a *theory*. Theories are sets of concepts that provide accepted, reliable, and systematic accounts of nature.

QUIZ

Fill in the blanks:

1. A solidifying agent made from red algae is known as: _____

2. A generalized approach used by many scientists to conduct research: _____

3. A _____ is an educated guess based on observations.

4. A means of testing a hypothesis is called a(an): _____

5. When a hypothesis survives testing and is accepted, it is called a(an): _____

Match the scientist:

    A. Eli Metchnikoff
    B. Theodore Schwann
    C. Alexander Fleming
    D. Walther Hesse
    E. Walter Hesse

6. Developed the cell theory. _____

7. Discovered that leukocytes engulf and digest microorganisms. _____

8. Discovered that *Penicillium notatum* produces the antibiotic penicillin. _____

9. Used agar as a solidifying agent. _____

10. Developed the germ theory. _____

The Chemical Elements of Microorganisms

In this chapter, you will explore the chemistry of microorganisms and examine the chemical processing that changes the chemistry of food into chemical elements that provide energy and nutrients to a microorganism.

CHAPTER OBJECTIVES

In this chapter, you will

- Learn about chemical elements
- Begin to understand chemical bonds
- Examine catalysts
- Become familiar with molarity

Microorganisms are composed of chemicals. In fact, chemicals make up all organisms, which is why understanding the chemical components of a microorganism is a fundamental component of learning microbiology.

A microorganism must ingest and digest food to live. *Food* is composed of molecules that microorganisms process into the chemical elements they need to sustain life. This is called the *digestive process*. Chemicals that are not needed by the microorganism are excreted as *waste*.

Atoms retained by the microorganism are rearranged to form new compounds that provide energy and nutrients the microorganism requires for movement, eating, and reproduction.

Heterotopic organisms are those that ingest and digest other organisms for energy and nutrients. *Autotrophic organisms* create their own food and do not need other microorganisms for survival.

Homeostasis is the state of equilibrium or balance that occurs within an organism when the organism is working normally. For example, when a pathogenic microorganism infects another organism, the chemical processing is disrupted. This occurs when you catch a cold or develop an infection. The pathogen hampers the organism by stopping or altering some chemical processing or metabolism in some way. Metabolism is the sum of all the chemical reactions that occur within the organism. The organism's immune system can typically counteract the disruption by the pathogen. Medication can also be taken to help restore the chemical processing. The goal is to return to homeostasis—normal cell function.

Everything Matters

Anything that takes up space and has mass is matter. The chair you're sitting on is matter. You are matter. And so are the microorganisms crawling over you and the chair. All nonliving and living things are matter because they take up space and have mass. It is easy to envision something taking up space, but what is mass?

Mass is the amount of matter a substance or an object contains. A common misconception is that mass is the *weight* of a substance. It is true that the more there is of a substance, the more it weighs. However, weight is the force of gravity acting on mass and is calculated as weight = mass × gravity or W=mg. A trip to the moon will clarify the difference between mass and weight: You have the same mass on Earth as you do on the moon, but you weigh more on the Earth than you do on the moon because the Earth has six times the gravitational force of the moon.

Chemical Elements and the Atom

Everything, including you, is composed of chemical elements. A *chemical element*, sometimes referred to simply as an *element*, is a substance that cannot be broken down into simpler substances by a chemical process. All matter is a combination of chemical elements.

A chemical element is made up of atoms. An *atom* is the smallest particle of an element; it cannot be decomposed further into smaller chemical substances (Fig. 2-1). In the early 1800s, John Dalton developed the *atomic theory*, which explains the relationship between an element and an atom. The atomic theory states that an element cannot be decomposed into two or more chemical substances because the element consists of only one kind of atom. The atom is also the smallest amount of matter that can enter into a chemical reaction. You'll learn about chemical reactions later in this chapter.

At the center of every atom is a *nucleus*. The nucleus does not change spontaneously unless it is unstable—making it radioactive—and does not participate in chemical reactions. It is for this reason that the nucleus for most atoms is considered stable. The nucleus is made up of protons and neutrons. A *proton* is a positively charged particle. A *neutron* is a particle that does not have a charge; it is called *neutral* or *uncharged*. The number of protons in the nucleus equals the number of electrons in an electrically stable atom.

Moving around the nucleus are *electrons*. An electron is a negatively charged particle that follows a path called an *orbital*. Electrons are the part of an atom that enters into chemical reactions.

This makes the atom neutral because the number of positively charged particles (protons) offsets the number of negatively charged particles (electrons).

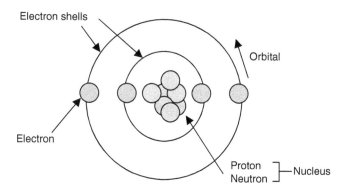

FIGURE 2-1 · An atom is the smallest particle of an element.

Protons and neutrons are made up of smaller particles called *hadrons*. Hadrons are formed from building-block particles called *quarks*. There are two types of hadrons: *baryons*, which are made up of three quarks, and *mesons*, which are made up of one quark and one antiquark. Electrons, on the other hand, are made up of *leptons*.

An element is identified by its atomic number. The *atomic number* is the number of protons in the nucleus of the atom. The *atomic mass* (also called the *atomic weight*) is slightly less than the sum of the masses of an atom's neutrons and protons. The standard for measuring atomic mass is called a *dalton*, named for John Dalton. A dalton is also known as an *atomic mass unit* (amu). For example, a neutron has an atomic mass of 1.088 daltons. A proton has an atomic mass of 1.077 daltons. An electron has an atomic mass of 0.0005 dalton.

Atoms that have the same atomic number are classified as the same chemical element because these atoms behave the same way. Therefore, a chemical element consists of atoms with the same number of protons.

Atoms of elements that have the same atomic number but different mass numbers are called *isotopes*. This difference is the result of a difference in the number of neutrons.

Each chemical element is identified by a symbol that is either the first letter or the first letter and another letter in its name. For example, the symbol C is used for carbon. Some chemical elements have English names, whereas others have Latin names. It is for this reason that symbols for some chemical elements seem strange at first glance. Take sodium, for example. You would think that its symbol should be S, but that's the symbol for sulfur. The symbol for sodium is Na—the first two letters of its Latin name, *natrium*.

There are 92 natural chemical elements and others that scientists have synthesized (created). All these are organized into a table called the *periodic table* (see "A Dinner Table of Elements: The Periodic Table" below). The six most abundant chemical elements in living organisms are carbon, oxygen, hydrogen, nitrogen, phosphorus, and calcium. Many others are also important and are found in trace amounts (Table 2-1).

A Dinner Table of Elements: The Periodic Table

As scientists continued to discover new chemical elements, it became apparent that a mechanism was needed to place chemical elements in some kind of order. In this way, scientists could easily reference information about each chemical element.

TABLE 2-1 Chemical Elements Commonly Found in All Living Things			
Element	Symbol	Atomic Number	Approximate Atomic Weight
Calcium	Ca	20	40
Carbon	C	6	12
Chlorine	Cl	17	35
Hydrogen	H	1	1
Iodine	I	53	127
Iron	Fe	26	56
Magnesium	Mg	12	24
Nitrogen	N	7	14
Oxygen	O	8	16
Phosphorus	P	15	31
Potassium	K	19	39
Sodium	Na	11	23
Sulfur	S	16	32

In the 1800s, Russian chemist Dmitri Mendeleev organized the known elements into a table by their atomic weights. Chemist H. G. J. Moseley reorganized the elements using their atomic numbers rather than their atomic weights. Elements were placed in the table based on increasing atomic number. This is referred to as the *law of chemical periodicity*, and the table became known as the *periodic table* (Fig. 2-2).

The periodic table consists of seven rows, each called a *period*. Chemical elements that have the same number of electron shells are placed in the same period. Rows are divided into columns, which are identified with the Roman numerals IA through VIIIA or the numbers 1 through 18 depending on the author of the periodic table. Chemical elements within the same column have the same chemical properties. For example, chemical elements in column IA can be joined easily with other chemical elements. In contrast, chemical elements in column VIIIA will not join with other chemical elements.

Each chemical element is identified by its symbol on the periodic table and is associated with two numbers. The number on top of the chemical symbol is the atomic number. The number beneath the chemical symbol is the atomic weight.

I A	II A	III A	IV A	V A	VI A	VII A	VIII A	VIII A	VIII A	I B	II B	III B	IV B	V B	VI B	VII B	VIII B
1 **H** 1.0079																	2 **He** 4.003
3 **Li** 6.94	4 **Be** 9.0121											5 **B** 10.81	6 **C** 12.011	7 **N** 14.006	8 **O** 15.999	9 **F** 18.998	10 **Ne** 20.17
11 **Na** 22.989	12 **Mg** 24.035											13 **Al** 26.981	14 **Si** 28.085	15 **P** 30.973	16 **S** 32.06	17 **Cl** 35.453	18 **Ar** 39.948
19 **K** 39.098	20 **Ca** 40.08	21 **Sc** 44.955	22 **Ti** 47.90	23 **V** 50.941	24 **Cr** 51.996	25 **Mn** 54.938	26 **Fe** 55.847	27 **Co** 58.933	28 **Ni** 58.71	29 **Cu** 63.546	30 **Zn** 65.38	31 **Ga** 69.735	32 **Ge** 72.59	33 **As** 74.921	34 **Se** 78.96	35 **Br** 79.904	36 **Kr** 83.80
37 **Rb** 85.467	38 **Sr** 87.62	39 **Y** 88.905	40 **Zr** 91.22	41 **Nb** 92.906	42 **Mo** 95.94	43 **Tc** 98.906	44 **Ru** 101.07	45 **Rh** 102.90	46 **Pd** 106.4	47 **Ag** 107.86	48 **Cd** 112.41	49 **In** 114.82	50 **Sn** 118.69	51 **Sb** 121.75	52 **Te** 127.60	53 **I** 126.90	54 **Xe** 131.30
55 **Cs** 132.90	56 **Ba** 137.33	57 **La** 138.90	72 **Hf** 178.49	73 **Ta** 180.94	74 **W** 183.85	75 **Re** 186.20	76 **Os** 190.2	77 **Ir** 192.22	78 **Pt** 195.09	79 **Au** 196.96	80 **Hg** 200.59	81 **Tl** 204.37	82 **Pb** 207.2	83 **Bi** 208.98	84 **Po** (209)	85 **At** (210)	86 **Rn** (222)
87 **Fr** (223)	88 **Ra** 226.02	89 **Ac** (227)	104 **Unq** (261)	105 **Unp** (262)	106 **Unh** (263)	107 **Uns** (262)	108 **Uno** (265)	109 **Une** (266)	110 **Unn** (272)								

Lanthanide Series

58 **Ce** 140.12	59 **Pr** 140.90	60 **Nd** 144.24	61 **Pm** (145)	62 **Sm** 150.4	63 **Eu** 151.96	64 **Gd** 157.25	65 **Tb** 158.92	66 **Dy** 162.5	67 **Ho** 164.93	68 **Er** 167.26	69 **Tm** 168.93	70 **Yb** 173.04	71 **Lu** 174.96

Actinide Series

90 **Th** 232.03	91 **Pa** 231.03	92 **U** 238.02	93 **Np** 237.04	94 **Pu** (244)	95 **Am** (243)	96 **Cm** (247)	97 **Bk** (247)	98 **Cf** (251)	99 **Es** (254)	100 **Fm** (257)	101 **Md** (258)	102 **No** (259)	103 **Lr** (260)

FIGURE 2-2 • The periodic table organizes chemical elements into periods and groups.

The Glowing Tale of Isotopes

Scientists describe the decay of an isotope using half-life. The *half-life* of an isotope is the time required for half the radioactive atoms in a sample of the isotope to decay into a more stable form. The rate at which the number of atoms of an isotope disintegrates is called the isotope's *rate of decay*, which can be a matter of seconds, minutes, hours, days, or years. Ernest Rutherford coined the term *half-life* at the turn of the twentieth century. Rutherford discovered two kinds of radiation that he called *alpha* and *beta* radiation. Scientists acknowledged Rutherford's important contribution by naming an element for him: rutherfordium (Rf).

Around the same time, Marie Curie, along with her husband, Pierre Curie, discovered that atoms of the chemical elements polonium (Po) and radium (Ra) decayed spontaneously and gave off particles. She called this process *radioactivity*.

A chemical element can have multiple isotopes. Each of those isotopes has the same atomic number but a different mass number. As you'll recall from earlier in this chapter, the mass number is the sum of protons and neutrons in the nucleus. Each isotope of the same chemical element has a different number of neutrons but the same number of protons.

Around They Go: Electronic Configuration

Previously in this chapter you learned that electrons of an atom move around the atom's nucleus in a pattern called an *orbital*. An atoms's orbitals are organized into one or more energy levels around the nucleus. The lowest energy level is the orbital closest to the nucleus. The highest energy level is the outermost orbital. The outermost orbital is called the *valence shell*. Each orbital holds a maximum number of electrons. The innermost, or first, shell holds a maximum of two electrons, the second can hold a maximum of eight, and the third can hold 18. Larger elements that contain many more shells can contain many more electrons. There can be as many as seven shells.

An atom with completely filled shells is called an *inert atom* and is said to be *chemically stable*. An inert atom tends not to react with other atoms. However, an atom that has an incomplete set of electrons in its valence shell is *chemically unstable* and tends to react with other atoms in an effort to become stable. Atoms want to be stable, so they either empty or fill their valence shells.

In order to become stable, the atom must undergo a chemical reaction to acquire one or more electrons from another atom, give up one or more electrons to another atom, or share one or more electrons with another atom.

A *chemical reaction* is the formation and breakdown of chemical bonds. All chemical reactions involve a shift of atoms from one molecule or ionic compound to another.

Chemical reactions occur naturally, sometimes taking a relatively long time to complete. A *catalyst* can be used to speed up a chemical reaction. Catalysts are molecules that speed up the rate of a reaction by lowering the energy of activation needed for the reaction to occur. *Enzymes* are molecules, typically proteins found in living cells that act like catalysts to increase the rate of reaction without changing the products of the reaction or by being consumed in the reaction. In summary, catalysts remain unaffected by the chemical reaction and do not affect the result of the reaction. A catalyst simply speeds up the reaction.

Before James, There Was Bond . . . Chemical Bond

An atom becomes stable by bonding with one or more atoms in order to fill its valence shell. When two atoms of the same chemical element bond together, they form a *diatonic molecule*. When two atoms of different elements bond, they form a chemical *compound*.

Two atoms are held together by the attractive force between them. Energy is required for the chemical reaction to bond atoms. This energy is termed potential chemical energy because it is stored in the resulting molecule or compound.

For example, combining two atoms of hydrogen forms a hydrogen molecule, H_2 (Fig. 2-3). Combing a hydrogen molecule consisting of two atoms with one oxygen atom forms the compound we know as water, H_2O (Fig. 2-4).

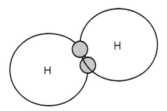

FIGURE 2-3 • Hydrogen becomes chemically stable by sharing a valence electron with another hydrogen atom.

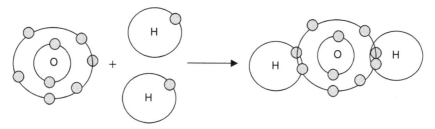

FIGURE 2-4 • Water is a compound consisting of two hydrogen atoms and one oxygen atom.

Bonds are formed in two ways:

- The atoms gain or lose an electron from the valence shell; this is called an *ionic attraction*.
- The atoms share one or more electrons in the valence shell; this is called a *covalent bond*.

In reality, atoms bond together using a range of ionic and covalent bonds. There are four kinds of chemical bonds:

- *Ionic bond*. This bond involves the transfer of electrons from one atom to another atom. An atom becomes unbalanced when it gains or loses an electron. An atom that gains an electron becomes negatively charged, and an atom that loses an electron becomes positively charged. An atom that is involved in this exchange is called an *ion*. The atom that gives up an electron is called a *cation*. A cation is positively charged. The atom that receives an electron is called an *anion*, which is negatively charged. The reaction that creates table salt from sodium and chlorine is due to an ionic bond between these atoms (Fig. 2-5).
- *Covalent bond*. This bond occurs when atoms share electrons in their valence shells (Fig. 2-5). The shared electron orbits the nucleus of both atoms. A covalent bond is the strongest bond and the one found most commonly in organisms. There are three kinds of covalent bonds: *single, double,* and *triple*. These names reflect the number of electron pairs that are shared between the two atoms that form the bond. Atoms that share electrons *equally* form *nonpolar covalent bonds*. Atoms that share electrons *unequally* form *polar covalent bonds*.
- *Coordinate covalent bond*. This bond is formed when electrons of the shared pair come from the same atom.

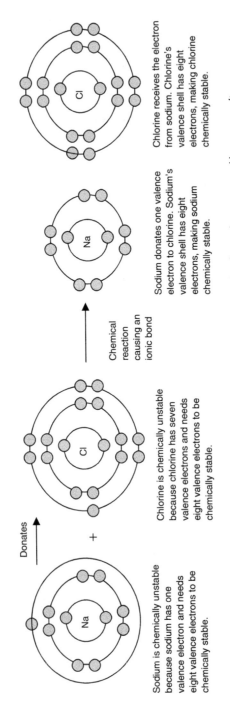

Sodium is chemically unstable because sodium has one valence electron and needs eight valence electrons to be chemically stable.

Chlorine is chemically unstable because chlorine has seven valence electrons and needs eight valence electrons to be chemically stable.

Sodium donates one valence electron to chlorine. Sodium's valence shell has eight electrons, making sodium chemically stable.

Chlorine receives the electron from sodium. Chlorine's valence shell has eight electrons, making chlorine chemically stable.

Donates

Chemical reaction causing an ionic bond

FIGURE 2-5 • Sodium donates one valence electron to chlorine in a chemical reaction that forms the compound known as *salt*.

- *Hydrogen bond.* A hydrogen bond forms a weak (5 percent of the strength of a covalent bond), temporary bond that serves as a bridge between either different molecules or portions of large macromolecules. For example, two water molecules are combined physically using a hydrogen bond.

Decoding Chemical Shorthand

Over the years, chemists have developed a way of describing atoms, chemical elements, and reactions so that they can convey ideas to each other. Table 2-2 shows commonly used chemical notations that you'll need to know when learning about microbiology.

I Just Want to See Your Reaction

The process of bonding atoms together and separating atoms that are already bonded together is called a *chemical reaction*. A chemical reaction is a process that involves the rearrangement of the structure of a substance. It can lead to

TABLE 2-2 Commonly Used Chemical Notation

Notation	Description
Na^+	The plus superscript indicates a positive ion.
Cl^-	The negative superscript indicates a negative ion.
$Na^+ + Cl^- \rightarrow NaCl$	The plus sign indicates synthesizing (combining) two ions. The right arrow indicates that a chemical reaction occurs toward the product.
$NaCl \rightarrow Na^+ + Cl^-$	Decomposing (breaking up) a molecule or chemical compound.
$NaOH + HCl \rightarrow NaCl + H_2O$	Exchange reaction where a chemical compound is decomposed into its chemical components and those components are synthesized into a new compound. Here, sodium hydroxide (NaOH) and hydrochloric acid (HCl) form salt (NaCl) and water (H_2O).
$Na^+ + Cl^- \leftrightarrows NaCl$	Reversible reaction is noted with a left arrow over a right arrow.
C—C	Single covalent bond.
C=C	Double covalent bond.
C≡C	Triple covalent bond.
H_2O	A subscript following a chemical symbol indicates the number of atoms (two hydrogen atoms). If no subscript is used, then it is implied that there is one atom (here, one oxygen atom).

$$X + Y \underset{\text{water}}{\overset{\text{heat}}{\rightleftharpoons}} XY$$

FIGURE 2-6 · In theory, all chemical reactions are reversible. In practice, these are called *reversible reactions*.

changes in structure and energy of atoms but does not affect the nucleus of the atoms.

For example, a chemical reaction occurs when a sodium atom is combined with a chlorine atom; the resulting chemical compound is table salt. If the sodium chloride (table salt) compound were broken down into its chemical components, you would see that the atoms of sodium and chlorine remain unchanged.

Theoretically, a chemical reaction can be reversed if the conditions are optimal. A chemical reaction that is reversible is called a *reversible reaction* (Fig. 2-6).

In practical use, some reactions can do this much more easily than others. Some of these reversible reactions occur owing to the instability of the reactants and products, whereas others will reverse only under special conditions. Examples of special conditions could be the presence of water or the application of heat.

There are three basic types of chemical reactions:

- *Synthesis reaction.* Two or more atoms, ions, or molecules are bound to form a larger molecule. A synthesis reaction combines substances called *reactants* to form a new molecule, called a *product*. Reactants are substances that interact in a reaction, and the *product* is the result of this reaction. In $Na^+ + Cl^- \rightarrow NaCl$, sodium and chlorine are reactants and sodium chloride is the product of this reaction. A synthesis reaction in a living organism is referred to as an *anabolic reaction*, or *anabolism. Anabolic reactions use energy (are endergonic).*

- *Decomposition reaction.* This is a reaction that breaks the bond between atoms in a molecule or chemical compound. In $NaCl \rightarrow Na^+ + Cl^-$, sodium chloride is broken up into its chemical elements, sodium and chlorine. A decomposition reaction in a living organism is called a *catabolic reaction*, or *catabolism. Catabolic reactions release energy (are exergonic).*

- *Exchange reaction.* This is a reaction that is both a synthesis reaction and a decomposition reaction, where a chemical compound is decomposed into its chemical components and then the atoms are rearranged into a new molecule or molecules. In $NaOH + HCl \rightarrow NaCl + H_2O$, sodium hydroxide ($NaOH$) and hydrochloric acid (HCl) enter into an exchange reaction to form salt ($NaCl$) and water (H_2O).

A chemical reaction theoretically can be reversed, but in practice, some reactions create unstable compounds that may require special conditions to exist for the reverse reaction to happen. The special conditions required to reverse a reaction appear below the arrow in the reaction notation. Above the arrow appears any special condition that must exist for the synthesized reaction to occur. In Fig. 2-6, a temperature of 250°C is the special condition for the synthesized reaction to occur, and absolute zero is necessary for the decomposition reaction to occur.

Still Struggling

It can be confusing understanding the difference between a *synthesis reaction* and a *decomposition reaction*. Think of it this way. If you wanted to create music, you could use a music *synthesizer*. Any time you create something, you *synthesize* it. If you wanted to build a house, you would take smaller material, such as wood, and nail it together to make the larger structure, the house. This act would cause you to use energy. The same thing happens when larger, more complex molecules are formed by putting together smaller atoms, ions, and molecules. This act also uses energy (the input of energy), and we call that use of energy *anabolism* or a *synthesis reaction*.

If something decomposes, on the other hand, it falls apart. Think of that same house you just built. If you were to blow up that big structure into a million pieces, it would give off a lot of energy (energy is released). We would call this release of energy *catabolism*, or a *decomposition reaction*. Large, complex chemical structures are being broken down into smaller molecules, ions, and atoms.

Catalysts: Making Things Happen

Living organisms possess large molecules of proteins called *enzymes*. These enzymes act as catalysts. A *catalyst* is a chemical substance that speeds up the rate of a chemical reaction. These catalysts do this without affecting the end products of the reaction or permanently altering themselves.

In order for an enzyme to be effective, it must interact with a chemical called a *substrate*. The enzyme attaches itself to the substrate in an area that most likely will increase its ability to react. This *enzyme-substrate complex* lowers the activation energy of the reaction and enables the collision of chemicals involved in the reaction to be more effective.

An important thing about enzymes is that they can reduce the reaction time without increasing temperature. This is very important in living organisms because high temperatures can break apart the proteins that make up a cell.

Molarity: Hey, There's a Mole among Us

It seems nearly impossible to measure a molecule's mass or size. Fortunately, there is *Avogadro's number*, which is the number of particles in a mole of a substance. The number is 6.022×10^{23}. Amedeo Avogadro was an Italian physicist for whom the value was named.

Scientists can measure molecules using units called a *mole*, abbreviated as *mol*. One mole is equal to the atomic weight of an element expressed in grams. A mole is the weight in grams of a substance that is equal to the sum of the atomic weights of the atoms in a molecule of the substance. This is referred to as the *gram molecular weight*.

Let's look at a water molecule to determine how many moles there are in a liter of water.

- Find the atomic mass for each element that makes up water. Water has two hydrogens and one oxygen.
- Look up the symbol for each element on the periodic table. These are H and O for hydrogen and oxygen.
- Note the number below the symbol. This is the atomic mass for the chemical element. These are 1 for hydrogen and 16 for oxygen.
- Multiply the number of atoms of each element in the molecule by its atomic mass to determine the value for one mole of the chemical element. For water, there are two hydrogen atoms, so this will be 2×1 g. One mole of a hydrogen molecule H_2 equals 2 g. Water has one oxygen atom. Therefore, multiply 1×16 g. One mole of oxygen is 16 g.
- Add the atomic masses of atoms that make up the molecule to determine one mole of the molecule. For water, this is 2 g + 16 g = 18 g. One mole of water equals the atomic mass of 18 g.

- The weight of one mole is the atomic mass of a molecule expressed in grams. Therefore, one mole of water weighs 18 grams.
- A liter of water has a mass of 1,000 grams. Calculate the number of moles per liter by dividing the number of grams (1,000) by one mole of water (18 grams). The result is 55 moles/liter.

An Unlikely Pair: Inorganic and Organic

Chemical compounds are divided into two general categories of substances. These are

- *Inorganic compounds.* These are compounds that do not contain the chemical element carbon (C).
- *Organic compounds.* These are compounds that contain carbon atoms. The exception is carbon dioxide (CO_2). Carbon dioxide is inorganic.

Inorganic compounds are further divided into three categories. These are

- *Acids.* An acid is any compound that dissociates into one or more hydrogen ions (H^+) and one or more negative ions (called *anions*) and is a proton donor.
- *Bases.* A base is any compound that dissociates into one or more positive ions (called *cations*) and one or more negative hydroxide ions. The negative hydroxide ions (OH^-) can either accept or share protons.
- *Salts.* A salt is an ionic compound that dissociates into one or more positive or negative ions in water, although some salts are not soluble in water. The positive and negative ions are neither hydrogen ions nor hydroxide ions. Sodium and chlorine atoms break away from the salt lattice when water molecules surround them. Water molecules become oriented so that the positive poles face the negatively charged chlorine ions and the negative poles face the positively charged sodium ions. The water's hydrogen shells react with the sodium and chlorine ions, disrupting the bonds between sodium and chlorine and dissolving it.

The pH Scale

There must be a balance between acids and bases to maintain life. An imbalance disrupts homeostasis. The acid-base balance is measured using the pH scale. The *pH scale* (Fig. 2-7) measures the acidity or alkalinity (base) of a

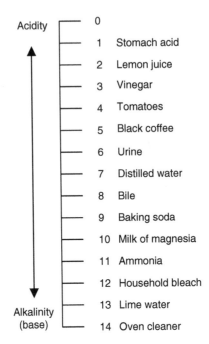

Acidity

	0	
	1	Stomach acid
	2	Lemon juice
	3	Vinegar
	4	Tomatoes
	5	Black coffee
	6	Urine
	7	Distilled water
	8	Bile
	9	Baking soda
	10	Milk of magnesia
	11	Ammonia
	12	Household bleach
	13	Lime water
	14	Oven cleaner

Alkalinity
(base)

FIGURE 2-7 • The pH scale is a logarithmic scale that measures the acidity or
alkalinity of a substance.

substance using a pH value from 0 to 14. Values on the pH scale are logarithmic. A pH value of 7 is neutral, which is the pH of pure water. A pH value
greater than 7 is a base or alkaline. A pH value of less than 7 is an acid. A change
in one pH value is a large change because it is a logarithmic scale. For example,
a pH of 1 has 10 times more hydrogen ions than a pH of 2 and 100 times more
hydrogen ions than a pH of 3 ($pH = \sim\log_{10}[H^+]$).

Adding a substance that will increase or decrease the concentration of hydrogen ions can change the pH value of a substance. Increasing the concentration
of hydrogen ions makes the substance more acidic, and decreasing the concentration makes the substance more alkaline.

The pH value of chemical compounds in living things naturally fluctuates
during metabolism. *Metabolism* is a collection of chemical reactions occurring
in a living organism. Sometimes the chemical compound is more acidic than
alkaline, and vice versa. Any drastic sway in the acid-base balance could have a
devastating effect. A chemical compound called a *buffer* is used to prevent
harmful swings in the acid-base balance. A buffer releases hydrogen ions or
binds hydrogen ions to stabilize the pH.

Organic Compounds

An *organic compound* is a compound whose chemical elements include carbon. Carbon plays an import role in living things because compounds that contain it build many different organic compounds, each having different structures and functions. The large size of most carbon-containing molecules and the fact that they don't dissolve easily in water make them useful in building body structures. Organic compounds also store energy required by an organism for metabolism.

Carbon can combine with other atoms because carbon has four electrons in its outer shell. This leaves room for four additional electrons from other atoms to bond to the carbon atom in a biological reaction (Fig. 2-8). Carbon also has low electronegativity and lacks polarity when a bond is formed.

A carbon atom commonly combines with other carbon atoms to form a *carbon chain*. There are two types of carbon chains—*straight carbon chains* and *ring carbon chains*. Figure 2-9 shows how a carbon ring is used to illustrate fructose. Carbon chains are the basic form for many organic compounds.

FIGURE 2-8 • The lines indicate a single bond with other atoms.

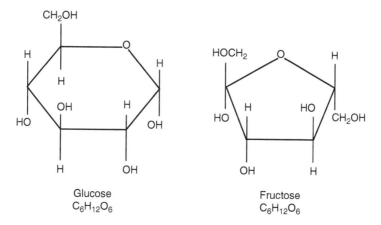

Glucose
$C_6H_{12}O_6$

Fructose
$C_6H_{12}O_6$

FIGURE 2-9 • A carbon ring chain is used to show the compound fructose.

Organic compounds come in many sizes—small to large. Many, but not all, large organic compounds are called *polymers*. A polymer is made up of small molecules called *monomers*. A monomer is another name for *subunit*. Monomers are bonded together to form a polymer in a process called *dehydration synthesis*, which removes water molecules from the compound.

A large organic compound is called a *macromolecule* and in many cases is a polymer. A *macromolecule* can be reduced to its monomer in a process called *hydrolysis*, which adds water molecules to the polymer.

Four types of organic compounds are macromolecules. These are

- Carbohydrates
- Lipids
- Proteins
- Nucleic acids

Carbohydrates

Carbohydrates store energy in an organism in the form of sugar, starches, and, in the human body, glucogen. *Cellulose* is also a carbohydrate used as bulk to move food and waste through the gastrointestinal tract. Carbohydrates are also used as material in the cell wall. Carbohydrates are organized into three major groups. These are

- *Monosaccharides.* Some of the important monosaccharides are glucose, the main energy source for an organism; fructose, acquired by eating fruit; galactose, which is in milk; deoxyribose, DNA; and ribose, RNA. A monomer is also a monosaccharide.
- *Disaccharides.* This is a combination of two monosaccharides bonded during dehydration synthesis. Sucrose (table sugar) and lactose (milk sugar) are disaccharides. Sucrose contains glucose and fructose. Lactose contains glucose and galactose.
- *Polysaccharides.* A polysaccharide is comprised of many monosaccharides and includes glycogen, starch, cellulose, and chitin, which is an amino sugar.

Lipids

Lipids are our fats and provide protection and insulation and can be used as an energy reserve. They are important components of the cell membrane and store pigments. Lipids are not polymers.

There are four kinds of lipids. These are

- *Triglycerides*. Triglycerides protect and insulate the body from most lipids and are a source of energy. Because lipids have few polar covalent bonds, they are mostly insoluble (do not mix well) with polar solvents such as water.
- *Phospholipids*. Phospholipids are a major component in cell membranes.
- *Steroids*. Steroids are derived from cholesterol. Examples of steroid hormones include testosterone and estrogen.
- *Eicosanoids*. Eicosanoids are divided into two kinds. These are prostaglandins and leukotrienes. Prostaglandins are involved in various behaviors, such as dilating airways, regulating body temperature, and aiding in the formation of blood clots. Leukotrienes are involved in inflammatory and allergic responses.

Other kinds of lipids include fatty acids, lipoproteins, many plant pigments, including chlorophyll and beta-carotene, and the fat-soluble vitamins A, D, E, and K.

Proteins

The hundreds of proteins in a single cell comprise about 50 percent of a cell's dry weight. Proteins are important components of cell structure and function. Some help transport other chemicals into and out of the cell and are called *transporter proteins*, while others, such as hemoglobin, transport substances throughout the body. Some proteins are the *enzymes* that speed up chemical reactions within the cell. Some proteins that are produced by bacteria kill other bacteria and are called *bacteriocins*. The exotoxins produced by some microorganisms that cause disease are proteins. Antibodies that fight infections are proteins. Proteins that make up connective tissue give our body structure, and proteins in muscle help the muscle to contract. Proteins are made up of long chains of amino acids held together by peptide bonds. There are four structural levels of proteins. These are

- *Primary*. The primary structure is the sequence in which amino acids are linked to form the polypeptide. Sequences are determined genetically, and even the slightest alteration within the sequence may have a dramatic effect on the way the protein functions.
- *Secondary*. The secondary structure is locally folded and is the repeated twisting of the polypeptide chain that links together the amino acids.

There are two types of secondary structures. These are a *helix* and a *pleated sheet*. The *alpha helix* is a clockwise spiral structure. The *pleated sheet* forms the parallel portion of the polypeptide chain.

- *Tertiary.* The tertiary structure is the three-dimensional active structure of the polypeptide chain. Tertiary structure is the minimal level of structure for biological activity. This is due to chemical interactions between the side chains of the 20 different amino acids.

- *Quarternary.* Refers to complex proteins composed of subunits of polypeptide chains. An example would be DNA polymerase or hemoglobin.

Still Struggling

Proteins have many roles in living organisms. Remember, in vertebrates (eukaryotes), some important proteins are: *myosin* (this contributes to muscle contraction), *actin* (this also contributes to muscle contraction), *collagen* (provides support and protection in bone and other connective tissue), *immunoglobulins* (which are antibodies that provide protection from infection), *hemoglobin* (this transports oxygen, O_2, and carbon dioxide, CO_2, in blood), and *enzymes*. (An enzyme that is a biological catalyst increases the rate of chemical reactions in cells by reducing the energy required to begin the reaction. The reaction does not change the enzyme. The name of an enzyme typically ends with *-ase*.)

In prokaryotic organisms proteins also serve as enzymes, give the organism structure, are used in signaling between adjacent cells, transport substances into and out of the cell, can be toxins, and are a component in the flagella (flagellin).

The Blueprint of Protein Synthesis

Proteins play a critical role in chemical reactions of microorganisms and other kinds of organisms. Information needed to direct the synthesis of a protein is contained in deoxyribonucleic acid (DNA). This information is transferred through generations from parent to progeny. Nucleotides also store energy in high-energy bonds and form together to make nucleic acids.

There are three parts to a nucleotide:

- A nitrogen base, such as adenine
- A five-carbon sugar, such as ribose
- One or more phosphate groups

Nucleic acids are large molecules made up of repeating nucleotides. Nucleic acids are long polymer chains that contain all the genetic material of the cell and are found in the nucleus of the cell in eukaryotic organisms and in the cytoplasm of prokaryotic organisms. This genetic material determines the activities of the cell and is passed on from generation to generation.

Types of Nucleic Acids

Two types of nucleic acids are found in the cell:

- *Deoxyribonucleic acid (DNA)*. DNA is a double strand of nucleotides that is wound together. A gene is a segment of DNA. These gene segments are composed of a continuous series of nucleotides. Genes determine the genetic markers that are inherited from previous generations of the organism. A *genetic marker* is a specific genetic characteristic or trait such as the ability to synthesize a specific protein. Protein controls activities of the cell. Some microorganisms, such as viruses, contain either DNA or RNA but not both. Think of DNA as a set of instructions.
- *Ribonucleic acid (RNA)*. RNA is a single strand of nucleotides that relays instructions from genes to ribosomes, guiding the chemical reactions in the synthesis of amino acids into protein. Think of RNA as the person who carries out the instructions of DNA.

The Powerhouse: ATP

Energy needed for cellular activity is stored in adenosine triphosphate (ATP) molecules. ATP is considered the "energy currency" or "money" of the cell. ATP supplies the power necessary to

- Move flagella in microorganisms
- Move chromosomes in the cytoplasm

- Transport substances into and out of the plasma membrane
- Power synthesis reactions

Energy is released from ATP when the last (third) phosphate group is split from the ATP molecule by a hydrolysis reaction. The result is adenosine diphosphate (ADP). Whenever more ATP is needed the reaction can go in the opposite direction. A phosphate group can attach to the ADP molecule to make ATP. The energy needed to attach this phosphate group to ADP is supplied mainly by the breakdown of glucose by glycolysis and the citric acid cycle.

Scientists and Their Contributions to the Science of Chemistry

Year	Scientist	Contribution
1803	John Dalton	Developed the atomic theory that explains the relationship between an element and an atom.
1860	Dmitri Mendeleev	Organized chemical elements into a table according to their atomic weights.
1800	H. G. J. Moseley	Organized chemical elements into a table according to their atomic numbers. The table was known originally as the *law of chemical periodicity* and has since been called the *periodic table*.
1911	Ernest Rutherford	Developed the Rutherford model of an atom and developed the concept of half-life.
1903	Marie and Pierre Curie	Discovered radioactivity.
1811	Amedeo Avogadro	An Italian physicist after whom the value of the number of molecules in a mole of a substance was named.
1913	Niels Bohr	Proposed that electrons occupy a cloud surrounding the nucleus of an atom. This is called an *orbital*.

QUIZ

Fill in the blanks:

1. The center of an atom is called the _____.

2. Negatively charged particles that orbit around the nucleus are called _____.

3. Charged particles of an atom that are positive are called _____.

4. Negatively charged particles follow a path called a(an) _____.

5. The number of protons in the nucleus is called the _____.

Match the scientist:

    A. John Dalton
    B. Dmitri Mendeleev
    C. Ernest Rutherford
    D. Marie Curie
    E. Niels Bohr

6. Organized chemical elements into a table according to their atomic weights. _____

7. Discovered radioactivity. _____

8. Proposed that electrons occupy a cloud surrounding the nucleus of an atom. _____

9. Developed the atomic theory. _____

10. Developed the concept of half-life. _____

chapter 3

Observing Microorganisms

Microorganisms are on our hands, arms, and everything we touch. You can see them with the aid of a microscope. In this chapter, you'll learn how to use a microscope to view microorganisms.

CHAPTER OBJECTIVES

In this chapter, you will

- Be introduced to the metric system
- Start measuring microorganisms
- Learn about magnification
- Become familiar with the various types of microscopes

The best defense against catching a cold is to wash your hands several times a day, especially before touching food. Health-care professionals wash their hands between seeing patients so that they avoid spreading disease. The term *spreading disease* is a more formal way of expressing what you were told growing up: Wash your hands so that you don't spread germs!

After reading this book you will realize that a *germ* is the common term used for a *microorganism*. The challenge for health-care professionals, teachers, and parents is to convince others to wash their hands even though they seem clean to the naked eye. When you look at your hands, you don't see microorganisms. Even with a magnifying glass, microorganisms are invisible.

Size Is a Matter of Metrics

We could describe the size of a microorganism in a variety of ways. A microorganism is very small. It is smaller than a human hair. Millions of them can fit on the head of a pin. All these descriptions give you an idea of how small a microorganism really is, but there's a problem using such words to describe size. What is tiny? Are we talking about the diameter of a human hair or its length? As for the head of a pin, how big is the pin?

Words that normally are used to describe size do so in relative terms rather than provide a precise measurement. Relative terms compare one thing with another thing without scientific precision. For example, the statement, "Millions of them can fit on the head of a pin" isn't precise and raises a lot of questions. How many millions? How big is the head of the pin?

Speaking in relative terms is fine if you want to convey a general sense of size. Saying that millions of microorganisms can fit on the head of a pin gives someone a sense that a microorganism isn't the size of a dog or cat but is something much smaller. However, scientists need to measure the size of a microorganism precisely to prevent diseases.

Let's see how this works by examining a surgical mask commonly used by medical professionals to control the spread of microorganisms. If you could zoom in on a surgical mask, you would see that the mask is a weave of threads that form a tiny web consisting of squarish holes. The size of each hole is determined by how close each strand of thread is to each other. Size is critical to reduce the spread of disease-carrying microorganisms. If the hole is smaller than the microorganism, then the microorganism is unable to pass through the surgical mask. It becomes trapped or simply moves in a direction of less resistance. However, a microorganism easily can pass through a hole that is larger than it is.

No Feet, Please

Scientists measure the size of microorganisms—and practically everything else—by using metric measurements, commonly called the *metric system*. A *system* is a way of doing something, such as having a system to beat the odds in Las Vegas. The *metric system* is a way of measuring things by using multiples or fractions of 10, called *factors of 10* or *the power of 10*.

The metric system is the standard way of measuring things throughout the world except in the United States, where we use the *U.S. Customary System of Measurement*, which includes inches, feet, and yards. The metric system is part of the *Système International d'Unités (SI system)*.

It is usually at this point in the study of microbiology that some students begin to slowly panic because they must learn a new measurement system. Don't panic! The metric system is very easy to learn—much easier to learn than the U.S. *Customary System of Measurement*.

The Prefix Fixes All Your Problems

The first trick to learning the metric system is to memorize the meaning of six prefixes. In the metric system, each prefix means that a meter is either multiplied or divided by a multiple of 10.

The second trick to learning the metric system is to learn how to multiply and divide by 10. This is easy because all you need to do is move the decimal point left or right. The decimal point is moved to the right one place when multiplying by 10. The decimal point is moved to the left one place when dividing by 10.

Let's see how this works. First, multiply 1 meter by 10.

$$1 \times 10 = 10$$

Now convert a meter to a decimeter. From the preceding section, you know that a decimeter is one-tenth of a meter, which means that you must divide by 10.

$$1 \div 10 = 0.10$$

Table 3-1 shows the prefixes for the metric system and their equivalents in meters.

A *meter* is the standard for length in the metric system. A *gram* is the standard for mass in the metric system. A *gram* uses the same prefixes as a meter to specify the number of grams that are represented by a value. For example, a kilometer is 1,000 meters and a kilogram is 1,000 grams. This makes it a lot easier to learn the metric system because the number of grams and meters are indicated by the same set of prefixes. The basic unit of volume is the liter.

TABLE 3-1 Prefixes	
Prefix	**Value in Meters**
Kilo (km) (kilo = 1,000)	1,000 m
Deci (dm) (deci = 1/10)	0.10 m
Centi (cm) (centi = 1/100)	0.01 m
Milli (mm) (milli = 1/1000)	0.001 m
Nano (nm) (nano = 1/1,000,000,000)	0.000000001 m
Pico (pm) (pico = 1/1,000,000,000,000)	0.000000000001 m

TABLE 3-2 Units of Mass in the Metric System	
Prefix	**Value in Grams**
Kilo (kg)	1,000 g
Hecto (hg)	100 g
Deka (dag)	10 g
Gram (g)	1 g
Deci (dg)	0.1 g
Centi (cg)	0.01 g
Milli (mg)	0.001 g
Micro (µg)	0.000001 g
Nano (ng)	0.000000001 g
Pico (pg)	0.000000000001 g

Table 3-2 contains a list of various ways to express a gram. You'll notice that this table contains three prefixes that were not used in Table 3-1. These are *deka-*, *hecto-* and *micro-*. The prefix *deka-* means 10, *hecto-* means 100, and *micro-* means 0.000001. That's 10 grams, 100 grams and 0.000001 gram.

Sizing Up Microorganisms

How small is a microorganism? You are probably asking yourself this question after learning the metric system of measurement. The answer depends on the kind of microorganism you are measuring. As you'll remember from Chapter 1, there are two general categories of microorganisms. These are prokaryotes and eukaryotes. Figure 3-1 illustrates the relative size of a microorganism when compared with other things.

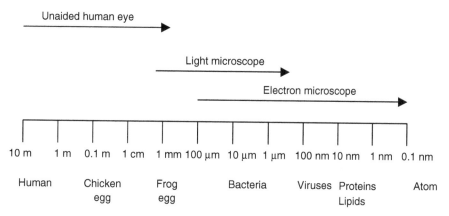

FIGURE 3-1 • Comparative sizes of humans and microorganisms.

Here's Looking at You

The principal way a microbiologist studies microorganisms is by observing them through a microscope. A *microscope* is a device that enlarges objects using a process called *magnification*. The simplest form of a microscope is a magnifying glass consisting of a single lens that is shaped in such a way as to make things appear larger than they are to the naked eye. And the simplest magnifying glass is the bottom of an empty glass. Some glasses are slightly bent at the bottom, causing a magnifying effect if the glass is held at a certain height over an object. This causes a change in the light ray's path.

The most complex microscope is an electron microscope, which uses electrons to bombard an object and ultimately magnify the object. Electron microscopes are capable of magnifying the organs of a microorganism called *organelles*, which you'll learn about later in this book. That is, an electron microscope is capable of showing what is inside a bacterium or virus.

Wavelength

You don't really "see" anything. It sounds strange, but it is true. You see only the reflection of light waves—or, in the case of an electron microscope, the reflection or absence of electrons. Electromagnetic radiation is generated by a variety of sources, such as the sun, a light bulb, or a radio transmitter. It takes the form of a wave similar in shape to an ocean wave.

A wave has two characteristics. These are the wave height and the wavelength. The *wave height* is the highest level above the surface traveled by the wave. Let's say that you're traveling across a calm stretch of ocean. This is the surface. Your boat then is pushed high above the surface by a swell—the wave height—before returning to the surface. The *wavelength* is the distance between the highest point of two waves. That is, the distance the boat travels between the highest point of the first wave and the highest point of the second wave.

Waves of electromagnetic radiation are in a continuous scale and are clustered into groups called *bands*. They are given names based on the length of the waves. Some are probably familiar to you, such as *x* rays, visible light waves, and radio waves. These groups are assembled into the *electromagnetic spectrum*.

Waves such as light waves are generated from a source such as the sun and strike an object such as your friend. Your friend absorbs some light waves and reflects other light waves. Your eyes detect only the reflected light waves.

It is this principle that enables you to observe a microorganism using a microscope. Light waves from a light bulb are reflected onto the microorganism. Reflected light waves are observed using the microscope. As you'll learn in Chapter 4, sometimes a microorganism reflects few light waves, making it difficult to see under a microscope. A *stain* is used to cause the microorganism to reflect different light waves. Microorganisms are visible under an electron microscope by directing waves of electrons onto the microorganism. Some electron waves are absorbed, and others are reflected. The reflected waves are detected by an electronic circuit that displays an image of the microorganism on a video screen.

What Big Eyes You Have: Magnification

Light reflected from a specimen travels in a straight line to your eyes, which lets you see the specimen at its natural size. You can magnify the size of the specimen by looking at the specimen through a lens.

The specimen is the *focal point*, which is the place where all the reflected light originates. Light travels in a straight line from the focal point to the lens, where the light is bent in a process called *refraction*. The angle at which light is bent is called the *angle of refraction*, which is measured in degrees from the natural path of the light. The degree of the *angle of refraction* is determined by the curvature in the lens. The more the lens curves, the greater is the angle of refraction.

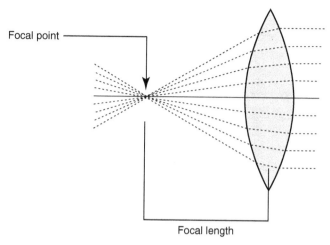

Focal point

Focal length

FIGURE 3-2 • The focal point is where the light rays meet.

The image appears larger as the light reflected from the image is refracted. Although the image appears magnified, curvature does distort the image. The amount of distortion depends on the angle of refraction and the distance between the lens and the specimen, which is called the *focal length*. The point at which light rays meet is called the *focal point* (Fig. 3-2).

You probably saw a distorted image when using a magnifying glass and were able to minimize the distorted effect by changing the distance between the magnifying glass and the specimen.

The Microscope

A *microscope* is a complex magnifying glass. In the 1600s, during the time of Antoni van Leeuwenhoek (see Chapter 1), microscopes consisted of one lens that was shaped so that the refracted light magnified a specimen 100 times its natural size. Other lenses were shaped to increase the magnification to 300 times.

However, van Leeuwenhoek realized that a single-lens microscope is difficult to focus. Once van Leeuwenhoek brought the specimen into focus, he kept his hands behind his back to avoid touching the microscope for fear that they would knock the microscope out of focus. It was common in the 1600s for scientists to make a new microscope for each specimen they wanted to study rather than to try to focus the microscope.

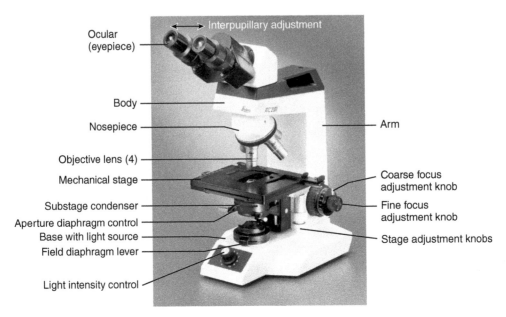

FIGURE 3-3 • Parts of a compound light microscope.

The single-lens microscope is a thing of the past. Scientists today use a microscope that has two sets of lenses (*objective* and *ocular*), which is called a *compound light microscope*. Figure 3-3 shows the parts of a compound light microscope. A compound light microscope consists of

- *Illuminator.* This is the light source located below the specimen.
- *Condenser.* A lens that focuses the light through the specimen.
- *Stage.* This is the platform that holds the specimen.
- *Objective.* This is the lens that is directly above the stage.
- *Nosepiece.* This is the portion of the body that holds the objectives over the stage.
- *Field diaphragm.* This controls the amount of light getting into the condenser.
- *Base.* This is the bottom of the microscope.
- *Coarse focusing knob.* This is used to make relatively wide focusing adjustments to the microscope.
- *Fine focusing knob.* This is used to make relatively small adjustments to the microscope.
- *Body.* This is the microscope body.

- *Ocular eyepiece.* This is the lens on the top of the body tube. It has a magnification of 10× normal vision.

Measuring Magnification

A compound microscope has two sets of lenses and uses light as the source of illumination. The light source is called an *illuminator* and passes light through a *condenser* and on through the specimen. Reflected light from the specimen is detected by the objective. The objective is designed to redirect the light waves, resulting in magnification of the specimen.

There are typically four objectives, each having a different magnification. These are 4×, 10×, 40×, and 100×. The number indicates by how many times the original size of a specimen is magnified, so the 4× objective magnifies the specimen four times the specimen's size. The eyepiece of the microscope is called the *ocular eyepiece*, and it, too, has a lens—called an *ocular lens*—that has a magnification of 10×.

You determine the total magnification used to observe a specimen under a microscope by multiplying the magnification of the objective by the magnification of the ocular lens. Suppose that you use the 4× objective to view a specimen. The image you see through the ocular is 40× because the magnification of the object is multiplied by the magnification of the ocular lens, which is 10×.

Many microscopes have several objectives connected to a revolving nosepiece above the stage. You can change the objective by rotating the nosepiece until the objective that you want to use is in line with the body of the microscope. You'll find the magnification marked on the objective. Sometimes the mark is color coded and other times the magnification is etched into the side of the objective.

Resolution

The area that you see through the ocular eyepiece is called the *field of view*. Depending on the total magnification and the size of the specimen, sometimes the entire field of view is filled with the image of the specimen. Other times, only a portion of the field of view contains the image of the specimen.

You've probably noticed that the specimen becomes blurry as you increase magnification. Here's what happens. The size of the field of view decreases as magnification increases, resulting in your seeing a smaller area of the specimen. However, the resolution of the image remains unchanged; therefore, you must adjust the fine focus knob to bring the image into focus again.

Resolution is the ability of the lens to distinguish fine detail of the specimen and is determined by the wavelength of light from the illuminator. At the beginning of this chapter, you learned about the wave cycle, which is the process of the wave going up and then falling down time and again. The *wavelength* is the distance between the peaks of two waves. As a general rule, shorter wavelengths produce higher resolutions of the image seen through the microscope.

Contrast

The image of a specimen must contrast with other objects in the field of view or with parts of the specimen itself to be visible in different degrees of brightness. Suppose that the specimen is a thin tissue layer of epidermis. The tissue must be a different color from the field of view; otherwise, the tissue and field of view blend, making it impossible to differentiate between the two. That is, the tissue and the field of view must contrast.

Previously in this chapter you learned that what you see is light reflected by the specimen (or the transmitted light if the specimen doesn't absorb light). The illuminator shines white light onto the specimen. White light contains all the light waves in the visible spectrum. The specimen absorbs some of the light waves and reflects other light waves, giving the appearance of some color other than white.

Light waves that are reflected by the specimen are measured by the refractive index. The *refractive index* specifies the number of light waves reflected by the object. There is a low contrast between a specimen and the field of view if they have nearly the same refractive index. The further these refractive indexes are from each other, the greater is the contrast between the specimen and the field of view.

Unfortunately, the refractive indexes of the specimen and the field of view are fixed. However, you can tweak the refractive index of the specimen by using a stain. The stain adheres to all or part of the specimen, absorbing additional light waves and increasing the difference between the refractive indexes of the specimen and the field of view. This results in an increase in the contrast between the specimen and the field of view.

Oil Immersion

A challenge facing microbiologists is how to maintain good resolution at magnifications of 100× and greater. To maintain good resolution, the lens must be small, and sufficient light must be reflected from both the specimen and the stain used on the specimen. The problem is that too much light is lost; air between the slide and the objective prevents some light waves from passing to the objective, causing the fuzzy appearance of the specimen in the ocular eyepiece.

The solution is to place a drop of oil onto the glass slide. The oil takes out the air, and because oil has the same refractive index as glass, the oil becomes part of the optics of the microscope. Light that is usually lost because of the air is no longer lost. The result is good resolution under high magnification.

Types of Light Compound Microscopes

There are three popular light compound microscopes used today (Table 3-3).

TABLE 3-3 Quick Guide to Microscopy		
Type of Microscope	**Features**	**Best Used for**
Bright-field microscope	Uses visible light	Observing dead stained specimens and living organisms with natural color
Dark-field microscope	Uses visible light from an illuminator located above the specimen that causes the rays of light to reflect off the specimen	Observing living, unstained organisms
Phase-contrast microscope	Uses a condenser that increases differences in the refractive index of structures within the specimen	Observing live specimens; allows for viewing internal structures of the specimen
Fluorescent microscope	Uses ultraviolet light to stimulate molecules of the specimen to make it stand out from its background	Observing specimens that are stained (tagged) with fluorescent antibodies
Electron Microscopes		
Transmission electron microscope	Uses electron beams and electromagnetic lenses to view thin slices of cells	Observing exterior surfaces and internal structures
Scanning electron microscope	Uses electron beams and electromagnetic lenses	Giving a three-dimensional view of exterior surfaces of cells

Bright-Field Microscope

The bright-field microscope is the most commonly used microscope and consists of three lenses. These are the ocular eyepiece and the objective. Light coming from the illuminator passes through the specimen. The specimen absorbs some light waves and passes along other light waves into the lens of the microscope, causing a contrast between the specimen and other objects in the field of view. Specimens that have pigments contrast with objects in the field of view and can be seen by using the bright-field microscope. Specimens with few or no pigments have a low contrast and cannot be easily seen with the bright-field microscope unless they are stained. Some bacteria have low contrast and so special stains are used to visualize them better.

Dark-Field Microscope

The dark-field microscope focuses the light from the illuminator onto the top of the specimen rather than from below the specimen. The specimen absorbs some light waves and reflects other light waves into the lens of the microscope. The field of view remains dark, whereas the specimen is illuminated, providing a stark contrast between the field of view and the specimen. Dark-Field Microscopes are often used for live, unstained specimens.

Phase-Contrast Microscope

The phase-contrast microscope bends light that passes through the specimen so that it contrasts with the surrounding medium. Bending the light is called *moving the light out of phase*. Since the phase-contrast microscope compensates for the refractive properties of the specimen, you don't need to stain the specimen to enhance the contrast of the specimen with the field of view. This microscope is ideal for observing living microorganisms that are prepared in wet-mounted slides so that you can study a living microorganism.

Fluorescent Microscope

Fluorescent microscopy uses ultraviolet light to illuminate specimens. Some organisms fluoresce naturally, that is, give off light of a certain color, when exposed to the light of a different color. Organisms that don't fluoresce naturally can be stained with fluorochrome dyes. When these organisms are placed under a fluorescent microscope with an ultraviolet light, they appear very brightly colored in front of a dark background.

Differential Interference Contrast Microscope (Nomanski)

The differential interference contrast microscope, which uses *Nomanski optics*, works in a similar way to the phase-contrast microscope. However, unlike the phase-contrast microscope (which produces a two-dimensional image of the specimen), the differential interference contrast microscope shows the specimen in three dimensions.

The Electron Microscope

A compound light microscope is a good tool for observing many kinds of microorganisms. However, it isn't capable of seeing the internal structures of a microorganism, nor can it be used to observe a virus. These are too small to effectively reflect visible light sufficient to be seen under a compound light microscope. To view the internal structures of viruses and microorganisms, microbiologists use an electron microscope, where specimens are viewed in a vacuum.

Developed in the 1930s, the electron microscope uses beams of electrons and magnetic lenses rather than light waves and optical lenses to view a specimen. Very thin slices of the specimen are cut so that the internal structures can be viewed. Microscopic photographs called *micrographs* are taken of the specimen and viewed on a video screen. Specimens can be viewed up to 200,000× normal vision. However, living specimens cannot be viewed because the specimen must be sliced.

Transmission Electron Microscope

The transmission electron microscope (TEM) has a total magnification of up to 200,000× and a resolution as fine as 7 nanometers. A nanometer is 1/1,000,000,000 of a meter. The transmission electron microscope generates an image of the specimen two ways. First, the image is displayed on a screen similar to that of a computer monitor. The image can also be displayed in the form of an electron micrograph, which is similar to a photograph. Specimens viewed by the TEM specimen must be cut into very thin slices; otherwise, the microscope does not depict the image adequately.

Scanning Electron Microscope

The scanning electron microscope (SEM) is less refined than the TEM. It can provide total magnification up to 10,000× and a resolution as close as 10 nanometers. However, a scanning electron microscope produces a three-dimensional image of a specimen. The specimen must be dehydrated and coated with a thin layer of gold, palladium, or other heavy metal.

Preparing Specimens

There are two ways to prepare a specimen to be observed under a compound light microscope. These are a *smear* and a *wet mount*.

Smear

A *smear* is a preparation process where a specimen is spread on a slide. You prepare a smear using the *heat fixation process*:

1. Use a clean glass slide.
2. Take a loop of the culture.
3. Place the live microorganism on the glass slide.
4. The slide is air dried and then passed over a Bunsen burner quickly three times.
5. The heat causes the microorganism to adhere to the glass slide. This is known as *fixing* the microorganism to the glass slide.
6. Stain the microorganism with an appropriate stain (see "Staining a Specimen" below).

Wet Mount

A *wet mount* is a preparation process where a live specimen in culture fluid is placed on a concave glass side or a plain glass slide. The concave portion of the glass slide forms a cuplike shape that is filled with a thick, syrupy substance such as *carboxymethyl cellulose*. The microorganism is free to move about within the fluid, although the viscosity of the substance slows its movement. This makes it easier for you to observe the microorganism. The specimen and the substance are protected from spillage and outside contamination by a glass coverslip that is placed over the concave portion of the slide.

Staining a Specimen

Not all specimens can be seen clearly under a microscope. Sometimes the specimen blends with other objects in the background because they absorb and reflect approximately the same light waves. You can enhance the appearance of a specimen by using a stain. A stain is used to contrast the specimen with the background.

A *stain* is a chemical that adheres to structures of the microorganism and in effect dyes the microorganism so that the microorganism can be seen easily under the microscope. Stains used in microbiology are either basic or acidic.

Basic stains are cationic and have a positive charge. Common basic stains are methylene blue, crystal violet, safranin, and malachite green. These are ideal for staining chromosomes and the cell membranes of many bacteria.

Acid stains are anionic and have a negative charge. Common acidic stains are eosin and picric acid. Acidic stains are used to stain cytoplasmic material and organelles or inclusions.

Types of Stains

There are two types of stains—*simple* and *differential*. Table 3-4 provides a summary of staining techniques.

Simple stain. A simple stain has a single basic dye that is used to show shapes of cells and the structures within a cell. Methylene blue, safranin, carbolfuchsin, and crystal violet are common simple stains that are found in most microbiology laboratories.

Differential stain. A differential stain consists of two or more dyes and is used in the procedure to differentiate between bacteria. Two of the most commonly used differential stains are Gram's stain and acid-fast stain.

Still Struggling

Sometimes staining techniques can be difficult to understand. Remember, examples of simple stains are methylene blue, safranin, and crystal violet, and these stains are used to highlight certain shapes of cellular structures and the arrangements of cells. Examples of differential stains include Gram's stain and acid-fast stains. These are used to differentiate bacteria. Special stains are used to distinguish structures such as endospores, capsules, and flagella.

In 1884, Hans Christian Gram, a Danish physician, developed Gram's stain. Gram's stain is a method for the differential staining of bacteria. Gram-positive microorganisms stain purple. Gram-negative microorganisms stain pink.

TABLE 3-4 Quick Guide for Staining Techniques

Type	Number of Dyes Used	Observations	Examples
Simple stains	Use a single dye	Size, shape, and arrangement of cells	Methylene blue, safranin, crystal violet
Differential stains	Use two or more dyes to distinguish different types or different structures of bacteria	Distinguish gram-positive or gram-negative	Gram's stain, Ziehl-Nielsen acid-fast stain
		Distinguish the members of *Mycobacteria* and *Nocardia* from other bacteria	
Special stains	Used to identify specialized structures	Exhibit the presence of flagella (Flagella stains)	Shaeffer-Fulton spore stain
		Exhibit endospores (Endospore stains)	
		Used to detect the presence of a capsule (negative stains); capsules usually do not stain, but look like a halo around the bacterial cell	

Staphylococcus aureus, a common bacterium that causes food poisoning, is gram-positive. *Escherichia coli* is gram-negative.

The *Ziehl-Nielsen acid-fast stain*, developed by Franz Ziehl and Friedrick Nielsen, uses a red dye called carcolfuchsin that attaches to the waxy material (mycolic acid) in the cell walls of bacteria such as *Mycobacterium tuberculosis*, which is the bacterium that causes tuberculosis, and *Mycobacterium leprae*, which is the bacterium that causes leprosy. Microorganisms that retain the red dye are called *acid-fast*. Those that do not retain this stain are counter stained with methylene blue. Bacterial cells that do not possess mycolic acid stain blue.

Here's how to stain a specimen with Gram's stain (Fig. 3-4):

1. Prepare the specimen using the heat fixation process (see "Smear" earlier in the chapter) and flood with crystal violet for one minute. This is a "primary stain."

2. Wash the slide front and back with water.

3. Apply iodine on the specimen using an eyedropper. The iodine helps the crystal violet stain to adhere to the specimen. Iodine is a *mordant*, which is a chemical that fixes the stain to the specimen.

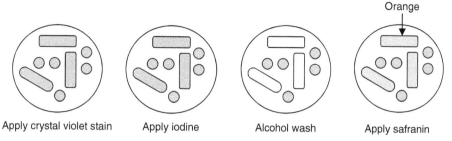

FIGURE 3-4 · How to stain a specimen with Gram's stain.

4. Wash the slide front and back with water. The specimen will now appear purple in color.

5. Wash the specimen with an ethanol or alcohol-acetone decolorizing solution until the alcohol runs clear.

6. Wash the specimen with water to remove the alcohol.

7. Apply the safranin stain (secondary stain or counter stain) to the specimen using an eyedropper.

8. Wash the specimen.

9. Use a paper towel to blot the specimen until it is dry.

10. The specimen is ready to be viewed under the microscope. Gram-positive bacteria appear purple, and gram-negative bacteria appear pink.

Here's how to apply the Ziehl-Nielsen acid-fast stain to a specimen:

1. Prepared the specimen (see "Smear" earlier in the chapter).

2. Apply the red dye carbolfuchsin stain generously using an eyedropper.

3. Let the specimen sit for a few minutes.

4. Warm the specimen over steaming water. The heat will cause the stain to penetrate the cell wall.

5. Wash the specimen with an alcohol-acetone decolorizing solution consisting of 3 percent hydrochloric acid and 95 percent ethanol. The hydrochloric acid will remove the color from non-acid-fast cells and the background. Acid-fast cells will stay red because the acid cannot penetrate the cell wall.

6. Apply methylene blue stain on the specimen using an eyedropper.

Special Stains

Special stains are paired with dye-specific structures of microorganisms such as endospores, flagella, and gelatinous capsules. For capsule staining, the bacteria is mixed in a solution containing either India ink or nigrosin to stain the background of the microorganism. A second stain like safran is used to stain the bacteria. Most dyes cannot stain capsules, they appear as halos around the bacteria that has been stained.

The *Schaeffer-Fulton endospore stain* is a special stain used to colorize endospores. The *endospore* is a dormant part within some bacterial cell that protects the bacterium from nonsuitable environments.

Here's how to apply the Schaeffer-Fulton endospore stain:

1. Prepare the specimen (see "Smear" earlier in the chapter).
2. Place malachite green on the specimen and steam over boiling water for 5–8 minutes. This allows the dye to penetrate the endospore.
3. Wash the specimen for 30 seconds.
4. Apply the safranin stain to the specimen using an eyedropper to stain parts of the cell other than the endospore.
5. Wash the slide.
6. Observe the specimen under a light microscope.

Scientists and Their Contributions

Year	Scientist	Contribution
1884	Hans Christian Gram	Developed Gram's stain used to stain and identify bacteria.
1882	Franz Ziehl and Friedrick Nielsen	Developed the Ziehl-Nielsen acid-fast stain used to stain bacteria.

QUIZ

Fill in the blanks:

1. This scientist developed Gram's stain to identify bacteria. _____

2. These two scientists developed the acid-fast stain. _____ and _____

3. The Schaeffer-Fulton stain is a(an) _____.

4. The amount of light waves reflected by an object is the _____.

5. The ability of a lens to distinguish fine detail of the specimen is the _____.

Match the microscope:

   A. Dark-field microscope
   B. Transmission electron microscope
   C. Scanning electron microscope
   D. Fluorescent microscope
   E. Bright-field microscope

6. Uses visible light for observing dead stained specimens and living organisms with natural color. _____

7. Focuses the light from an illuminator onto the specimen rather than from under the specimen. _____

8. Uses ultraviolet light to stimulate molecules of the specimen to make it stand out from its background. _____

9. Uses electromagnetic lenses to view exterior surfaces and internal structures of specimens. _____

10. Uses electromagnetic lenses to give a three-dimensional view of exterior surfaces of cells. _____

chapter 4

Prokaryotic Cells and Eukaryotic Cells

In this chapter, you'll learn about the six processes of life and about how the life processes are used by prokaryotic and eukaryotic cells.

CHAPTER OBJECTIVES

In this chapter, you will

- Learn the differences between prokaryotic and eukaryotic organisms and the differences in their structures
- Be introduced to the fluid mosaic model
- Become familiar with the cytoplasmic membrane

You are alive, and so are microorganisms, but the chair you are sitting on is not. You know this, but do you know what distinguishes living organisms from nonliving material? It does not depend on having a brain or a heartbeat, instead, all living organisms are alive because they carry on six processes unique to life.

The six processes of life focus on transforming nutrients into energy so that the organism can grow and reproduce to continue the life cycle of the species. You do this, and so does every microorganism, except that you can't see the microorganisms without a microscope.

Six Processes of Life

What do you and athlete's foot have in common? You'll recall from Chapter 1 that *tinea pedis* is the scientific name for athlete's foot and that it is caused by the *Trichophyton rubrum* fungus. Both of you are alive. Sometimes it's hard to imagine that microorganisms are alive because we can't see them with the naked eye—although they can make their presence known to us in annoying ways.

The six life processes require a living organism to

- *Undergo metabolization*. Metabolism involves the breakdown of nutrients for energy or the extraction of energy from the environment.
- *Respond to stimuli*. React to internal and external environmental changes.
- *Move*. This includes movement of the entire organism, of organs or organelles within that organism, or of individual cells that make up the organism.
- *Grow*. Increase in the size or number of cells.
- *Differentiate*. This ability allows cells that are unspecialized (have no specific function) to develop a specialized function which ultimately develops into trillions of cells with specific functions. The process where cells that are unspecialized become specialized. (An example would be a single fertilized human egg developing into an individual.) Prokaryotic cells do not differentiate.
- *Reproduce*. Form new cells to create a new organism.

For additional information, see Table 4-1.

TABLE 4-1 Basic Life Processes in Microorganisms			
Process	**Eukaryotic Cells**	**Prokaryotic Cells**	**Viruses**
Metabolism: Sum of all chemical reactions	Yes	Yes	Uses their host's cells for metabolism
Responsiveness: Ability to react to environmental stimuli	Yes	Yes	Do not respond to stimuli
			Attach to host cell receptors, which initiates the replication of a new virus
Movement: Motion of individual organelles, a single cell, or entire organism	Yes	Yes	Virions (viruses outside of a host cell) are non-motile
			Undergo Brownian movement (random collision)
Growth: Increase in size	Yes	Yes	No
Differentiate: When a cell that is unspecialized becomes specialized	Yes	No	No
Reproduction: Increase in number of cells or development of new organisms	Yes	Yes	No
			Viruses multiply, they do not reproduce

In this chapter, you will learn about cellular structure by exploring two kinds of cells: *prokaryotic* and *eukaryotic*. Bacteria are prokaryotic organisms. Animals, plants, algae, fungi, and protozoa are eukaryotic organisms. Table 4-2 provides a summary of general differences between prokaryotic and eukaryotic cells.

Prokaryotic Cells

Prokaryotic cells are very small, simple structures, and many have similar morphology, although there are many variations owing to the many differences in their genetic makeup and ecology. A *prokaryotic cell* is a cell that does not have a true nucleus. The nuclear region is called a *nucleoid*. The nucleoid contains most of the cell's genetic material and is usually a single circular molecule of DNA. *Karyo* is Greek for "kernel." A prokaryotic organism, such as a bacterium,

TABLE 4-2 Differences between Prokaryotic and Eukaryotic Cells

Characteristics	Prokaryotic Cells	Eukaryotic Cells
Cell wall	Typically contain peptidoglycan	Chemical composition varies
Plasma membrane	No carbohydrates	Contain carbohydrates
	No sterols	Contain sterols
Glycocalyx	Possess a capsule or a slime layer	Found in cells that lack a cell wall
Flagella	Cellular appendage	Extension of cell membrane
	Major protein is flagellin	Arrangement of microtubules
Cytoplasm	Actin cytoskeleton	Cytoskeleton primarily composed of actin and microtubules
		Undergoes cytoplasmic streaming
Membrane-bound organelles	None	Endoplasmic reticulum
		Golgi complex
		Lysosomes
		Peroxisomes
		Nucleus
		Mitochondria
		Chloroplasts
Ribosomes	70S	80S
		Ribosomes located in mitochondria and chloroplasts are 70S
Nucleus	No nuclear membrane	Have a nucleus surrounded by a nuclear envelope containing one or more nucleoli
	No nucleoli	
Chromosomes	Single circular chromosome	Multiple linear chromosomes
	No histones	Have histones
Cell division	Binary fission	Mitosis
Sexual reproduction	No meiosis	Meiosis
	DNA transferred in fragments	

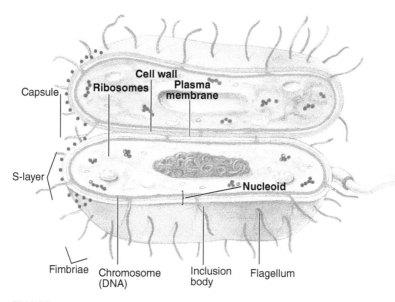

FIGURE 4-1 • A prokaryotic cell.

is a cell that lacks a membrane-bound nucleus or membrane-bound organelles. The exterior of the cell usually has a glycocalyx, flagella, fimbriae, and pili (Fig. 4-1).

Shapes of Prokaryotic Cells

Bacteria exist literally everywhere. The majority play a beneficial role, but some are pathogens that can cause disease. In this section we will focus on the shapes of bacteria that are pathogenic.

Bacteria can have a round shape, and such bacteria are called *cocci* or *coccus* (singular). These round forms of bacteria can occur in pairs, called *diplococci* or *diplococcus* (singular); in long chains, called *streptococcus*; or in clusters that look like a bunch of grapes, called *staphylococcus*.

An example of a diplococcus would be *Neisseria gonorrhoeae*. This is the organism that causes the sexually transmitted disease gonorrhea (Fig. 4-2).

An example of a chain of cocci would be *Enterococcus faecalis*, a pathogenic organism that can cause urinary tract infections and endocarditis (Fig. 4-3).

An example of staphylococcus would be *Staphylococcus aureus*, the pathogenic organism that can cause the skin condition impetigo (Fig. 4-4).

Rods are another bacterial shape. Such bacteria are commonly called *bacilli* or *bacillus* (singular). An example would be *Bacillus cereus*, the pathogenic

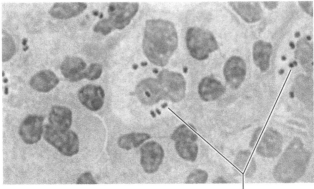

Diplococci

FIGURE 4-2 • *N. gonorrhoeae.* The diplococci are often within white blood cells (×1,000). (From Prescott et al., *Microbiology,* 6th ed., McGraw-Hill, 1996.)

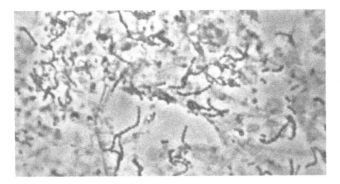

FIGURE 4-3 • *E. faecalis.* Note the chains of cocci; phase contrast (×200). (From Prescott et al., *Microbiology,* 6th ed., McGraw-Hill, 1996.)

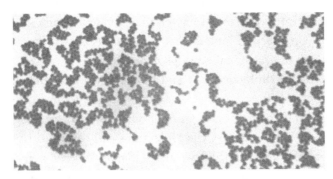

FIGURE 4-4 • *S. aureus.* Note the gram-positive spheres in irregular clusters; gram stain (×1,000). (From Prescott et al., *Microbiology,* 6th ed., McGraw-Hill, 1996.)

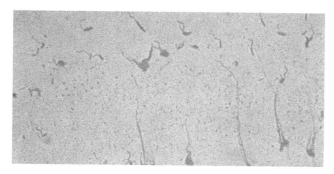

FIGURE 4-5 • *V. cholera.* Curved rods with polar flagella (×1,000). (From Prescott et al., *Microbiology*, 6th ed., McGraw-Hill, 1996.)

organism that can cause food poisoning. Some rods are curved and are called *vibrios.* An example of a curved-shaped bacteria is *Vibrio cholerae.* This pathogenic organism causes cholera (Fig. 4-5).

Other shapes of bacteria include filamentous, fungus-like bacteria which have *hyphae* and can branch into a cluster called *mycelium.* An example includes *Actinomyces israelii,* which causes actinomycosis, abscesses of the jaw and mouth (Fig. 4-6).

Bacteria can also take a helical form. *Spirochetes* form spirals or twisted *flexible* rods. If the rods are *rigid,* they are called *spirilla.* An example of a spirochete

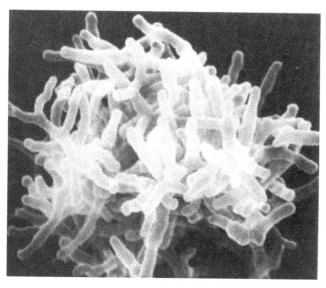

FIGURE 4-6 • *Actinomyces,* SEM (×21,000). (From Prescott et al., *Microbiology,* 6th ed., McGraw-Hill, 1996.)

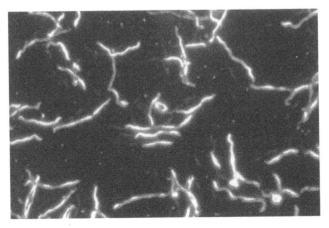

FIGURE 4-7 · *T. pallidum*, the spirochete that causes syphilis; dark-field microscopy (×500). (From Prescott et al., *Microbiology*, 6th ed., McGraw-Hill, 1996.)

would be *Treponema pallidum*, the organism that causes the sexually transmitted disease syphilis (Fig. 4-7).

An example of a spirillum would be *Spirillum volutans*, a nonpathogenic organism found in stagnant fresh water (Fig. 4-8).

Another odd-shaped type of bacteria is *Gallionella ferruginea*. This organism is found in areas containing ferrous iron, such as drainage bogs and iron springs (Fig. 4-9).

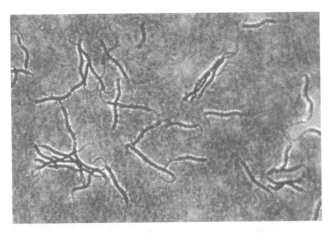

FIGURE 4-8 · *S. volutans*, a very large bacterium with flagellar bundles; phase-contrast microscopy (×210). (From Prescott et al., *Microbiology*, 6th ed., McGraw-Hill, 1996.)

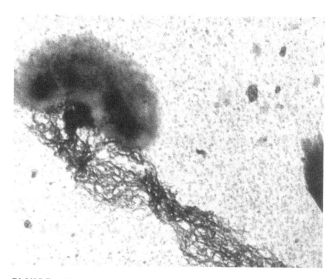

FIGURE 4-9 · *G. ferruginea* with stalk. (From Prescott et al., Micro-biology, 6th ed., McGraw-Hill, 1996.)

Prokaryotic Cell Structure

Glycocalyx

The *glycocalyx* is a sticky envelope, which is composed of polysaccharides and/or polypeptides, that surrounds the cell. The glycocalyx is found in one of two states. When it is firmly attached to the cell's surface it is called a *capsule*. When the glycocalyx is loosely attached to the cell wall it is called a *slime layer*. A slime layer is water-soluble and is used by a prokaryotic cell to adhere to surfaces external to the cell.

The capsule is used by some prokaryotic cells to protect against attack from the body's immune system. This is the case with *Streptococcus mutans*, which is a bacterium that colonizes teeth and excretes acid that causes tooth decay. Normally, the body's immune system would surround the bacterium and eventually kill it, but this doesn't happen with *S. mutans*. It has a protective capsule that prevents it from being recognized as a foreign microorganism by the body's immune system.

Flagella

Flagella are made of protein and appear whiplike (Fig. 4-10). They are used by the prokaryotic cell for motility. Flagella propel the microorganism away from harm and toward food in a movement known as *taxis*. Movement also occurs

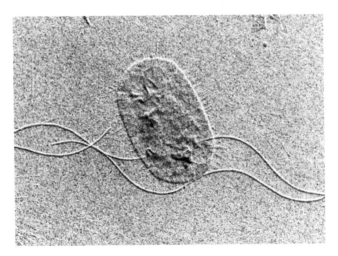

FIGURE 4-10 • A flagellum has a long tail that extends from the cell.

in response to a light or chemical stimulus. Movement caused by a light stimulus is referred to as *phototaxis*, and a chemical stimulus causes chemotactic movement to occur.

Bacterial cells can have single flagella, multiple flagella, or no flagella at all. The following terms refer to the position and number of flagella on a bacterium:

- *Monotrichous.* One flagellum.
- *Lophotrichous.* A clump of flagella, called a *tuft*, at one or both ends of the cell.
- *Amphitrichous.* Flagella at two ends of the cell.
- *Peritrichous.* Flagella covering the entire cell.
- *Endoflagellum.* A type of amphitrichous flagellum that is tightly wrapped around spirochetes. (A *spirochete* is a spiral-shaped bacterium that moves in a corkscrew motion. *Borrelia burgdorferi*, which is the bacterium that causes Lyme disease, possesses an endoflagellum.)

Fimbriae

Fimbriae are proteinaceous, sticky, bristle-like projections used by bacterial cells to attach to each other and to objects around them. *N. gonorrhoeae*, the bacterium that causes gonorrhea, uses fimbriae to adhere to the mucosal membrane of the genitourinary tract and to cluster cells of the bacteria.

Pili

Pili are tubules that are used to transfer DNA from one cell to another cell, similar to the tubes used to fuel aircraft in flight. Some are also used to attach one cell to another cell. The tubules are made of protein and are shorter in length than flagella and longer than fimbriae.

Cell Wall

The bacterial cell wall is located outside the plasma membrane and gives the cell its shape and provides rigid structural support. The cell wall also protects the cell from its environment.

As fluid containing nutrients enters the cell, pressure within the cell builds. It is the job of the cell wall to resist this pressure the same way that the walls of a balloon resist the buildup of pressure when the balloon is inflated. If pressure inside the cell becomes too great, the cell wall bursts, which is referred to as *lysis*.

The cell wall of many bacteria contains peptidoglycan and covers the entire surface of the cell. *Peptidoglycan* is made up of a combination of peptide fragments and carbohydrates, either *N*-acetylmuramic acid, commonly referred to as *NAM*, or N-acetylglucosamine, which is known as *NAG*.

Most microorganisms are colorless when viewed under a light microscope. In order to observe the structures of these organisms, we have to color stain them. A type of staining is called Gram staining, developed by Hans Gram. This type of differential staining is very useful because it classifies bacteria in either gram-positive or gram-negative groups.

- *Gram-positive.* Gram-positive bacteria retain the crystal violet stain after decolorization by alcohol. These bacteria appear purple when viewed under the microscope. The cell walls of gram-positive bacteria (Fig. 4-11) consist of peptidoglycan and teichoic acid.
- *Gram-negative.* Gram-negative bacteria lose the crystal violet color after decoloration by alcohol and become colorless. A dye called safranin is added which turns the bacteria pink when viewed under the microscope. The cell walls of gram-negative bacteria (Fig. 4-12) have a thin layer of peptidoglycan surrounded by a lipopolysaccharide outer membrane.

Cytoplasmic Membrane

The prokaryotic cell has a *cytoplasmic membrane* (Fig. 4-13) that forms the outer structure of the cell underneath the cell wall and separates the cell's internal structure from the environment. The cytoplasmic membrane is a membrane

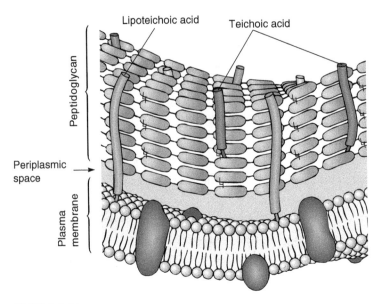

FIGURE 4-11 • Gram-positive cell wall.

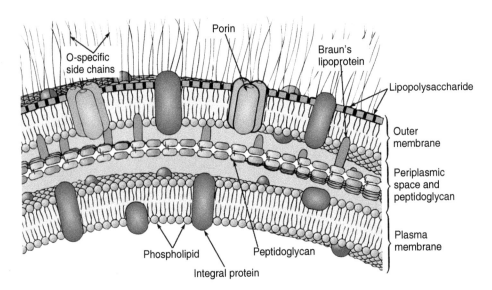

FIGURE 4-12 • Gram-negative cell wall.

that provides a selective barrier between the environment and the cell's internal structures and allows certain substances and chemicals to move into and out of the cell. The cytoplasmic membrane is composed of a bilayer of phospholipids. Phospholipids have a polar "head" and nonpolar "tails." This is referred to

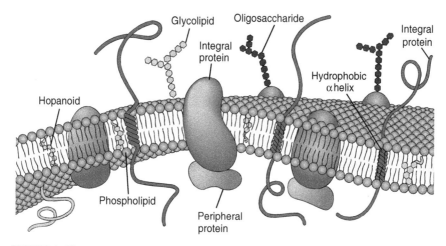

FIGURE 4-13 · The cytoplasmic membrane enables some substances to pass into and out of the cell.

as being *amphipathic*. Each polar part has a head that contains phosphate and is hydrophilic ("water loving"). Each nonpolar part has two tails composed of long fatty acids that are hydrophobic ("water fearing").

The heads always face an aqueous environment such as the extracellular fluid on the outside of the cell and the intracellular fluid inside the cell. Tails align back to back, preventing the fluid from crossing the cytoplasmic membrane.

The Fluid Mosaic Model

In 1972, S. J. Singer and G. L. Nicolson developed the *fluid mosaic model*, which describes the structure of the cytoplasmic membrane. *Mosaic* refers to the fact that proteins within the cytoplasmic membrane are arranged like tiles in a mosaic artwork. The term *fluid* implies that membrane proteins and lipids move freely within the cytoplasmic membrane. There are two kinds of proteins within the cytoplasmic membrane. These are

- *Integral proteins.* An integral protein spans across the phospholipid bilayer. Integral proteins typically are glycoproteins with a carbohydrate facing the exterior of the cell. They act like a molecular signature that cells use to recognize each other. Two examples are
  - *Carrier protein.* A transport protein that regulates the movement of specific molecules through the cytoplasmic membrane.

○ *Channel protein.* A channel protein forms pores or channels in the cytoplasmic membrane that permit the flow of ions through the cytoplasmic membrane. Other examples of integral proteins include receptors.

• *Peripheral proteins.* Peripheral proteins are on the inner and outer surfaces of the cytoplasmic membrane.

The Function of the Cytoplasmic Membrane

The cytoplasmic membrane regulates the flow of molecules (such as nutrients and gases) into the cell and removes waste from the cell. This involves various mechanisms, such as simple diffusion, facilitative diffusion, and active transport. In photosynthetic prokaryotes, the cytoplasmic membrane functions in energy production by collecting energy in the form of light.

The cytoplasmic membrane is *selectively permeable* because it permits the transport of some substances and inhibits the transport of other substances. Two types of transport mechanisms are used to move substances through the cytoplasmic membrane. These are passive transport and active transport.

Passive Transport

Passive transport moves substances in and out of the cell down a concentration gradient. Molecules move from an area of high concentration to an area of low concentration. There are three types of passive transport. They are

• *Simple diffusion.* Simple diffusion is the movement of substances from a higher-concentration region to a lower-concentration region (net movement). Only small chemicals (such as oxygen and carbon dioxide) or lipid-soluble chemicals (such as fatty acids) diffuse freely between the phospholipids of the cytoplasmic membrane using simple diffusion. Large molecules cannot fit between the phospholipids.

• *Facilitated diffusion.* Facilitated diffusion is the movement of substances from a higher-concentration region to a lower-concentration region (net movement) with the assistance of an integral protein. These "carrier" proteins are very specific for the molecules they transport. Once a carrier picks up a molecule such as glucose, it changes its shape and consequently releases the molecule on the opposite side of the membrane.

• *Osmosis.* Osmosis is the net movement (diffusion) of water from a region of higher water concentration to a region of lower concentration.

How Osmosis Works

Osmosis is a vital activity for cells. Depending on the water content of a cell compared with its environment, a cell can gain or lose water or stay the same. The following terms describe these conditions. Each term refers to the concentration of solute in the environment surrounding the cell.

- *Isotonic solution.* *Iso-* means "equal" if a cell is placed in an isotonic solution. This means that there is the same concentration of solute and solvent (water) inside and outside the cell. There is an *equal* movement of water into and out of the cell.

- *Hypertonic solution.* In a hypertonic solution, the cell is placed in an environment where there is a higher concentration of solute outside the cell than inside the cell. Consequently, the water inside the cell moves out of the cell by osmosis, causing the cell to shrink. This shrinking of the cell is called *crenation.* In cells possessing a cell wall it is called plasmolysis.

- *Hypotonic solution.* When a cell is placed in a hypotonic solution, there is more water outside the cell than inside the cell. This means that there is more solute concentration inside the cell and so the water outside will move into the cell by osmosis. This causes the cell to swell and ultimately burst. This is called *lysis.*

Active Transport

Active transport is the movement of a substance across the cytoplasmic membrane against the gradient using energy provided by the cell. This is similar to pumping water against gravity through a pipe. Energy must be spent for the pump to work.

A cell makes energy available by removing a phosphate (P) from adenosine triphosphate (ATP). ATP stores potential energy in its chemical bonds and is released by a chemical reaction within the cell. This energy is used to change the shape of the integral membrane protein—enabling substances to be pumped across the cytoplasmic membrane.

Group Translocation

Group translocation is a process that immediately modifies a substance once the substance passes across the cytoplasmic membrane. The cell must expend energy during group translocation, which is supplied by high-energy phosphate compounds such as phosphoenolpyruvic acid (PEP). Group translocation occurs in prokaryotic cells.

Endocytosis and Exocytosis

Endocytosis and exocytosis are processes used to move large substances or many small ones into and out of a cell. Large substances enter the cell by *endocytosis*. There are two kinds of endocytosis. These are phagocytosis and pinocytosis. *Phagocytosis* engulfs solid substances (large molecules), whereas *pinocytosis* engulfs liquid substances (small molecules). *Exocytosis* is the process that cells use to remove large substances, which is the way waste products and useful material such as hormones and neurotransmitters are secreted from a cell.

Cytosol and Cytoplasm

The *cytosol* is the intracellular fluid of a prokaryotic cell that contains proteins, lipids, enzymes, ions, waste, and small molecules dissolved in water. It is commonly referred to as being *semifluid*. Substances dissolved in the cytosol are involved in cell metabolism.

The cytosol also contains a region called the *nucleoid*, which is where the DNA of the cell is located. Unlike human cells, a prokaryotic microorganism has a single circular chromosome that isn't contained within a nuclear membrane or envelope.

The cytosol is located in the cytoplasm of the cell. *Cytoplasm* also contains the cytoskeleton, ribosomes, and inclusions.

Ribosomes

Ribosomes synthesize polypeptide. There are thousands of ribosomes in a cell. You'll notice them as the grainy appearance of the cell when viewing the cell with an electron microscope.

A ribosome is comprised of subunits consisting of protein and *ribosomal RNA* (rRNA). Ribosomes and their subunits are identified by their sedimentation rate. *Sedimentation rate* is the rate at which ribosomes are pulled to the bottom of a test tube by gravitational force when spun in a centrifuge. The sedimentation rate is expressed in *Svedberg units* (S). A sedimentation rate reflects the mass, size, and shape of ribosomal subunits. It is for this reason that the sedimentation rates of subunits of a ribosome do not add up to the ribosome's sedimentation rate.

Ribosomes in prokaryotic cells are uniquely identified by the number of proteins and rRNA molecules contained in the ribosome and by the sedimentation rate. Prokaryotic ribosomes are relatively small and less dense than ribosomes of other microorganisms. For example, bacterial ribosomes have a sedimentation rate of 70S compared with the 80S sedimentation rate of a eukaryotic ribosome, which you'll learn about later in this chapter.

Ribosomes and their subunits are targets for certain antibiotics that kill a bacterium by inhibiting the bacterium's protein synthesis. These antibiotics only kill cells that have a specific ribosome sedimentation rate. Cells with a different ribosome sedimentation rate are unaffected by the antibiotic. This enables the antibiotic to kill the bacteria and not the body that is infected by the bacteria.

Inclusions

An *inclusion* is a storage area that serves as a reserve for lipids, nitrogen, phosphate, starch, and sulfur within the cytoplasm. Scientists use inclusions to identify types of bacteria. Inclusions are usually classified as *granules*.

- *Granule inclusion.* Membrane-free and densely packed, this type of inclusion has many granules composed of specific substances. For example, *polyphosphate granules*, also known by the names *metachromatic granules* and *volutin*, have granules of polyphosphate that are used to synthesize ATP and are involved in other metabolic processes. A polyphosphate granule appears red under a microscope when stained with methylene blue.

- *Vesicle inclusion.* Surrounded by protein, this inclusion is commonly found in aquatic photosynthetic bacteria and cyanobacteria such as *phytoplankton*, which suspends freely in water. These bacteria use vesicle inclusions to store gas that gives the cell buoyancy to float at a depth where light, carbon dioxide (CO_2), and nutrients—all required for photosynthesis—are available.

Prokaryotic Size

Bacteria vary in size just as they do in shape. Some are as small as large viruses. An example would be *Mycoplasma pneumoniae*, the organism that causes walking pneumonia. The most common cause of bacterial pneumonia is *Streptococcus pneumonia*.

Other bacteria are relatively large. For example, *Epulopiseium fishelsoni* (Fig. 4-14) is about a million times larger than *Escherichia coli*.

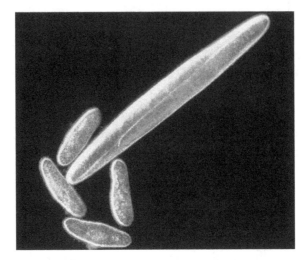

FIGURE 4-14 • This photograph, taken with pseudo dark-field illumination, shows *Epulopiseium fishelsoni* at the top of the figure dwarfing the paramecia at the bottom (×200). (From Prescott et al., *Microbiology*, 6th ed., McGraw-Hill, 1996.)

Eukaryotic Cells

A *eukaryotic cell* (Fig. 4-15) is larger and more complex than a prokaryotic cell and is found in animals, plants, algae, fungi, and protozoa. When you look at a eukaryotic cell under a microscope, you'll notice a highly organized structure

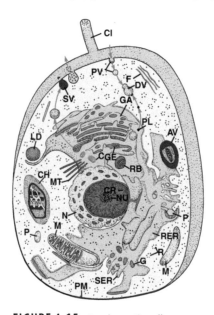

FIGURE 4-15 • A eukaryotic cell.

of organelles that are bound by a membrane. Each organelle performs a specialized function for the cell's metabolism. Eukaryotic cells also contain a membrane-bound nucleus where the cell's DNA is organized into linear chromosomes.

Depending on the organism, a eukaryotic cell may contain external projections of the cell membrane called *flagella* and *cilia*. These projections are used for moving substances along the cell's surface or for moving the entire cell. Flagella move the cell in a wavelike motion within their environment. *Cilia* move substances along the cell's surface and also aid in movement of some cells. Flagella and cilia are comprised of axoneme microtubules. An *axoneme microtubule* is a long, hollow tube made of protein called *tubulin*.

Still Struggling

Remember that eukaryotic cells have a nucleus surrounded by a nuclear membrane, contain membrane-bound organelles, and divide typically by mitosis. Prokaryotic organisms have no nuclear membrane or membrane bound organelles and divide by binary fission.

Cell Wall

Many eukaryotic cells have a cell wall. The composition of the cell wall differs with each organism. For example, the cell wall of many fungi is composed primarily of chitin. *Chitin* is a polysaccharide; it is a polymer consisting of *N*-acetylglucosamine (NAG) units. The polysaccharide cellulose is found in the cell walls of plants and many algae.

In contrast, protozoa have no cell wall and may have a pellicle. A *pellicle* is a flexible proteinaceous covering. Eukaryotic cells of other organisms (such as animals) that lack a cell wall have an outer plasma membrane that serves as an outside cover for the cell. All cells have a plasma membrane. The outer plasma membrane of some cells has a sticky carbohydrate called a *glycocalyx* on its surface. The glycocalyx is made up of covalently bonded lipids and proteins that form from glycolipid and glycoprotein in the plasma membrane. *Glycolipids* and *glycoproteins* anchor the glycocalyx to the cell, giving the cell strength and helping the cell to adhere to other cells. The glycocalyx is also a molecular signature used to identify cells from each other. White blood cells use the glycocalyx to identify a foreign cell before destroying it.

Plasma Membrane

The *plasma membrane* is a selectively permeable membrane enclosing the cytoplasm of a cell. This is the outer layer in animal cells. Other organisms have a cell wall as the outer layer, and the plasma membrane is between the cell wall and the cell's cytoplasm. The cell wall is the outer covering of most bacteria, algae, fungi, and plant cells. In eubacteria, which are prokaryotic microorganisms, the cell wall contains peptidoglycan.

The plasma (or cytoplasmic) membrane surrounds a eukaryotic cell and serves as a barrier between the inner cell and its environment. The *cytoplasmic membrane* is composed primarily of proteins and lipids. Exterior carbohydrates are used to uniquely identify the cell to other cells. Lipids known as *sterols* help to prevent the destruction of the cell when there is an increase in temperature, and they are used mainly for stability. *Lysis* is the destruction of a cell.

The plasma membrane of a eukaryotic cell functions like the plasma membrane of a prokaryotic cell, which you learned about previously in this chapter. That is, substances enter and leave a cell through the cytoplasmic membrane by using simple diffusion, facilitated diffusion, osmosis, or active transport.

Eukaryotic cells have the ability to change shape by extending parts or sections of the plasma membrane filled with cytoplasm. These extensions of the plasma membrane are called *pseudopods*. The word *pseudopod* means "false foot," and these "feet" enable the cell to have amoeboid motion. *Amoeboid motion* consists of muscle-like contractions that move the cell over a surface. Pseudopods are used to engulf substances and bring them into the cell, which is called *endocytosis* (a type of active transport). There are two types of endocytosis. These are *phagocytosis* (eat) and *pinocytosis* (drink). In phagocytosis, solid particles are engulfed by the cell. An example occurs when a white blood cell engulfs and destroys a bacterial cell. In pinocytosis, liquid particles are brought into the cell. An example occurs when extracellular fluid containing a substance is destroyed by the cell.

Cytoplasm and Nucleus

The cytoplasm of a eukaryotic cell contains cytosol, organelles, and inclusions, which is similar to the cytoplasm of the prokaryotic cell. Eukaryotic cytoplasm, as well as prokaryotic cytoplasm, contains a cytoskeleton that gives structure and shape to the cell and assists in transporting substances throughout the cell.

The nucleus of a eukarytoic cell contains DNA (hereditary information) and is contained within a nuclear envelope. DNA is also found in the mitochondria and chloroplasts. Depending on the organism, there can be one or more nucleoli within the nuclear envelope. A *nucleolus* ("little nucleus") is the site of ribosomal RNA synthesis, which is necessary for ribosomes to function properly.

In the nucleus, the DNA is wrapped around histone protein to enhance DNA packaging in the nucleus. When a eukaryotic cell is not in the reproduction phase, the DNA and its proteins look like a threaded mass called *chromatin*. When the cell goes through nuclear division, the strands of chromatin condense and coil together, producing rod-shaped bodies called *chromosomes*.

A eukaryotic cell uses a method of cell division during reproduction called *mitosis*. This is the formation of two genetically identical daughter cells from a parent cell.

Endoplasmic Reticulum

The *endoplasmic reticulum* contributes to the mechanical support and distribution of the cytoplasm and is the pathway for transporting lipids and some proteins throughout the cell. The endoplasmic reticulum also provides the surface area for the chemical reaction that synthesizes lipids.

The endoplasmic reticulum consists of *cisterns*, which are a network of flattened membranous sacs. The ends of these cisterns can be pinched off to produce membrane-enclosed sacs called *secretory vesicles*. Vesicles transport synthesized material in the cell.

There are two kinds of endoplasmic reticula:

- *Rough endoplasmic reticulum.* Covered by ribosomes, this is the site for synthesizing protein.
- *Smooth endoplasmic reticulum.* Not covered by ribosomes, this is the site for synthesizing lipids and steroids.

Golgi Complex

The Golgi complex is considered the "FedEx system" of the cell because it packages and delivers proteins, lipids, and enzymes throughout the cell and to the environment. The Golgi complex contains multiple cisternae stacked on top of each other like dinner plates. A *cisternae* is a sac or vessel that is filled with proteins or lipids to be packaged as they leave the Golgi complex and are transported to another part of the cell.

Lysosome

A *lysosome* is an organelle in animal cells that is formed by but is separate from the Golgi complex. It contains enzymes used to digest molecules that have entered the cell as well as disabled organelles within the cell. Think of lysosomes as the digestive system of the cell. For example, lysosomes in a white blood cell digest bacteria that are ingested by the cell during phagocytosis.

Mitochondrion

The *mitochondrion* is an organelle that is comprised of a series of folds (an outer membrane, an inner membrane space, and an inner membrane) called *cristae* that is responsible for the majority of a cell's energy production and cellular respiration. The Krebs cycle occurs within the center of the mitrochondrion, called the *matrix*; it is filled with *semifluid* in which adenosine triphosphate (ATP) is produced. ATP is the energy molecule in the cell. The mitochondrion is the powerhouse of the cell.

Chloroplast

Eukaryotic cells of green plants and algae contain *plastids*, one of which is the chloroplast. *Chloroplasts* are organelles that contain pigments of chlorophyll, carotenoids, and other light-absorbing molecules used for gathering light as well as enzymes necessary for photosynthesis. *Photosynthesis* is the process that converts light energy into chemical energy. The pigments are stored in membranous sacs called *thylakoids* that are arranged in stacks called *grana*.

Centriole

A *centriole* is one of a pair of cylindrical structures near the nucleus that is comprised of microtubules and aids in the formation of flagella and cilia. The centriole also has a part in eukaryotic cell division.

QUIZ

Fill in the blanks:

1. The ability to react to internal and external stimuli is called _____.

2. To increase the number of cells and/or create a new individual is called _____.

3. Motion of organelles of the entire organism is called _____.

4. Increase in the size and/or the number of cells is called _____.

5. The sum of the chemical reactions in an organism is called _____.

Match the organelle:

   A. Flagella
   B. Pili
   C. Fimbriae
   D. Glycocalyx
   E. Cell wall

6. Sticky envelope made of polysaccharides and/or polypeptides that surround the cell. _____

7. Used to propel the organism. _____

8. Composed of peptidoglycan in many bacteria. _____

9. Tubules that are used to transfer DNA from one bacterial cell to another. _____

10. Bristle-like projections used by bacterial cells to attach themselves to objects. _____

chapter 5

Chemical Metabolism

Metabolism is the collection of biochemical reactions that occur in our bodies and that begin when the digestive system breaks down food into biochemical components that our cells use for energy. In this chapter you'll learn about metabolism within the cell.

CHAPTER OBJECTIVES

In this chapter, you will

- Be introduced to metabolism
- Learn about catabolic and anabolic reactions
- Become familiar with enzyme activities
- Begin to understand the Krebs cycle

Cellular Metabolism

You probably have a relative such as Aunt Mary who cannot lose weight no matter what she does. She has been on every diet known to humankind and has signed up for all the exercise programs as far as the eye can see, yet she can't drop more than 5 pounds. Aunt Mary blames her metabolism as the underlying cause of her weight. She claims that her body processes food in the slow lane.

Whether the underlying cause of Aunt Mary's inability to lose weight is the speed of her metabolism or the coffee cake that she eats after every meal is still under investigation. However, metabolism does play an important role in all living organisms.

Riding the Metabolism Cycle

The components of a cell, including its plasma membrane and cell wall (and organelles in eukaryotic organisms), are made up of macromolecules that are linked together. The macromolecules are assembled from building blocks called *precursor metabolites*. Think of precursor metabolites as the bricks that are used to build a wall, and think of the wall as the macromolecule. Using the energy from adenosine triphosphate (ATP), these precursor metabolites are used to synthesize or build larger molecules. ATP is the short-term energy-storage molecule of the cell. Think of it as the battery pack of the cell. Cells use energy from ATP and enzymes to connect smaller molecules to form macromolecules. The cell grows as macromolecules are linked together and continue to grow into cellular structures such as organelles, plasma membranes, and cell walls.

Catabolic and Anabolic: The Only Reactions You Need

A biochemical reaction is called a *metabolic reaction*. Metabolic reactions fall into one of two classifications. These are catabolic reactions (catabolism) and anabolic reactions (anabolism).

A *catabolic reaction* is a metabolic reaction that releases energy as large molecules are broken down (metabolized) into smaller molecules. An example is when triglycerides and diglycerides are metabolized into glycerol and fatty acids.

An *anabolic reaction* requires energy as small molecules are combined to form larger molecules. This type of reaction is called *endergonic* because it uses free energy. For example, an anabolic reaction is the synthesis of proteins from amino acids.

A Little Give and Take: Oxidation-Reduction

Metabolic reactions sometimes involve the transfer of electrons from one molecule to another. One molecule *donates* an electron, and another molecule *accepts* the electron. This transfer of electrons is called *oxidation-reduction* or a *redox reaction*. A redox reaction consists of two events. The first event happens when a molecule donates an electron. This is called *oxidation*. The second event happens when another molecule accepts the donated electron. This is called *reduction*.

The cell uses *electron carrier molecules* to carry electrons between areas within the cell. Think of these carrier molecules as "shuttle buses." Carrier molecules are necessary because the cytoplasm of the cell does not contain free electrons.

Three important electron carrier molecules that are used in cell metabolism are

- Nicotinamide adenine dinucleotide (NAD^+)
- Flavine adenine dinucleotide (FAD^+)
- Nicotinamide adenine dinucleotide phosphate ($NADP^+$)

For example, when synthesizing ATP, NAD^+ carries electrons of a hydrogen (H) atom, making NADH. FAD carries two electrons of hydrogen, making $FADH_2$. Very often electrons of hydrogen atoms are the electrons transported by the carrier molecule. $NADP^+$ is used to reduce carbon dioxide (CO_2) to carbohydrates during the light-independent reactions.

Making Power: ATP Production

When enzymes break down nutrients (larger molecules) into smaller molecules, energy is released and can be used for future anabolic reactions. There are three mechanisms for ATP formation.

1. *Substrate-level phosphorylation.* Phosphate is transferred from another phosphorylated organic compound to adenosine diphosphate (ADP) to make ATP during an exergonic reaction.

2. *Oxidative phosphorylation.* Energy from the redox reactions of biochemical respiration is used to attach an inorganic phosphate to ADP to make ATP.

3. *Photophosphorylation.* Energy from sunlight is used to phosphorylate ADP with inorganic phosphate.

Remember: ATP cannot be stored. It is used immediately after synthesis. Consequently, ATP must constantly be made or the cell will die.

What's Your Name: Naming and Classifying Enzymes

Enzymes are typically named according to the substrate on which they act, and most end in the suffix *-ase*. Enzymes are classified into six major groups based on their actions. These classifications are

- *Hydrolases.* Enzymes in the *hydrolase* group increase a catabolic reaction by introducing water into the reaction. This reaction is called *hydrolysis*. For example, *lipase* (lipid + ase) is a hydrolytic enzyme that is used to break down lipid molecules.
- *Isomerases.* Enzymes in the *isomerase* group rearrange atoms within the substrate rather than add or subtract anything from the reaction. Phosphoglucoisomerase is an example of an isomerase because it converts glucose-6-phosphate into fructose-6-phosphate during the breakdown of glucose.
- *Ligases.* These are enzymes involved in anabolic reactions. These enzymes join molecules together and use energy in the form of ATP. An example is DNA ligase which is used during the synthesis of DNA.
- *Lyases.* Enzymes in the *lyases* group split molecules without using water in a catabolic reaction. For example, 1,6-biphosphate aldolase splits fructose-1,6-biphosphate into glucose-3-phosphate and dihydroxyacetone phosphate (DHAP) during glycolysis. These are catabolic reactions.
- *Oxidoreductases.* Enzymes in the *oxidoreductase* group oxidize (remove) electrons or reduce (add) electrons to a substrate in both catabolic and anabolic reactions. An example is lactic acid dehydrogenase, which oxidizes pyruvate to form lactic acid during fermentation.
- *Transferases.* Enzymes in the *transferase* group transfer functional groups from one molecule to another substrate in an anabolic reaction. A functional group could be a phosphate, an amino group, a carboxyl group, or a carbonyl group. For example, hexokinase transfers a phosphate group from ATP to glucose in the first step in the breakdown of glucose during the process of gycolysis.

Brewing Up Proteins

Most enzymes are proteins that can be inactive or active. An *inactive enzyme* does not act as a catalyst to increase the speed of a metabolic reaction. An *active enzyme* is a catalyst. An inactive enzyme is composed of *apoenzyme*; when an apoenzyme binds to its cofactor, the enzyme becomes active and is called a *holoenzyme*.

A *cofactor* is a substance that is either an inorganic ion, such as iron, magnesium, or zinc, or an organic molecule. Organic cofactors are called *coenzymes*. A coenzyme is a molecule that is required for metabolism. NAD+, NADP+, and FAD+ are examples of coenzymes. Some vitamins are coenzyme precursors.

The Magic of Enzymes: Enzyme Activities

All chemical reactions, including those which occur in metabolism, need a boost of energy to get started. The energy needed to begin a chemical reaction is called *activation energy*. An enzyme catalyzes a reaction by lowering the activation energy. Heat can lower the activation energy and set off a reaction. However, the temperature would be so high that the cell would die before the activation energy threshold could be reached.

Enzymes are needed for metabolism to occur in a timely fashion. The activity of enzymes depends on how closely their functional sites fit with their substrates. A substrate is a chemical that the enzyme reacts with, although the enzymes are not consumed in the reaction. The shape of the enzyme's functional site is called its *active site*. This site fits in regard to the shape of the substrate. The active site of the enzyme complements the shape of its substrate. A perfect fit *does not* occur until the substrate and enzyme bind together to form an enzyme-substrate complex.

The Right Influences: Factors Affecting Enzymes

The ability of an enzyme to lower the activation required for metabolism is influenced by three factors. These are pH, temperature, and the concentrations of enzyme, substrate, and product.

pH

The chemical denaturing of enzymes is caused by very high or very low pH. Hydrogen$^+$ ions that are released from acids and accepted by bases interfere with hydrogen bonding. If we change the pH of the environment of unwanted microorganisms, we can control their growth by denaturing their proteins. An example is vinegar, which is acetic acid; it has a pH of 3.0. Vinegar acts as a preservative in "pickling" vegetables. Ammonia has a pH of 11. Ammonia is a base, and for this reason, we use ammonia as a cleaner and disinfectant.

Enzyme Substrate Concentration

As substrate concentrations increase, enzyme activity also increases. When all enzyme binding sites have bound to a substrate, the enzyme has reached its *saturation point*. If more substrate is added, the rate of enzyme activity *will not* increase. One way organisms regulate their metabolism is by controlling the quantity and timing of enzyme synthesis.

Temperature

Changes in temperature change the shape of the active site and therefore influence the fit between the active site of the enzyme and the substrate. Enzymes in humans work best at about 37°C. This is the same temperature at which enzymes work best for most human pathogens. Once the temperature reaches the point that radically changes the shape of the active site, the bond between the active site and the substrate is broken, and this makes the enzyme ineffective. This is called *thermal denaturation*. Denatured enzymes lose their specific three-dimensional shape, making them nonfunctional. For example, the clear liquid portion of an egg turns to a white solid when the egg is heated. The clear liquid is made up of proteins. Heating these proteins denatures them.

Inhibitors

There are substances that block active sites of enzymes from bonding with a substrate. These substances are called *inhibitors*. There are two kinds of inhibitors: competitive and noncompetitive. A *competitive inhibitor* is a substance that binds to the active site of an enzyme, thus preventing the active site from binding with the substrate. For example, sulfa drugs contain the chemical sulfanilamide. Sulfa drugs inhibit microbial growth by fitting into the active site of an enzyme required in the conversion of *para*-aminobenzoic acid (PABA) into the B vitamin folic acid. Folic acid is needed for DNA synthesis in bacteria and thus prevents bacteria from growing. A noncompetitive inhibitor binds to another site on the enzyme called the *allosteric site* and in so doing alters the shape of the active site of the enzyme. The shape of the active site no longer complements the corresponding site on the substrate, and, therefore, no binding occurs. Noncompetitive inhibitors do not bind to active sites.

Carbohydrate Metabolism

Carbohydrates are the main energy source for metabolic reactions, and glucose is the most highly used carbohydrate in metabolism. Energy is produced by breaking down (catabolism) glucose in a process called *glycolysis*, which takes place in the cytoplasm of most cells. Glycolysis, also known as the *Embden-Meyerhof pathway* (Fig. 5-1), is the oxidation of glucose to pyruvic acid. In glycolysis, which originated from the Greek word *glykys* meaning "sweet" and *lysein* meaning "loosen," enzymes split a six-carbon sugar into two three-carbon sugars, which are then oxidized. Oxidation releases energy and rearranges the

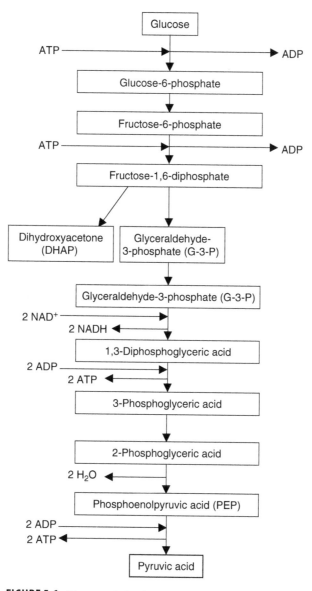

FIGURE 5-1 · Diagram of glycolysis: Embden-Meyerhof pathway.

atoms to form two molecules of pyruvic acid. It is during this process that NAD^+ is reduced to NADH with a net production of two ATP molecules.

Pyruvic Acid

Before entering the Krebs cycle, the pyruvic acid produced from the break-down of glucose must be further processed by converting it to *acetyl CoA*. This is accomplished by the enzyme complex *pyruvate dehydrogenase*. A carbon is removed from pyruvic acid as CO_2, and the product is an acetyl group (a two-carbon group). The acetyl group is attached to CoA, and the product is called *acetyl-CoA*. The removal of CO_2 is called *decarboxylation*.

Still Struggling

Remember, carbohydrates are the main energy source for metabolic reactions, and glucose is the most used carbohydrate in metabolism. The purpose of the Embden-Meyerhof pathway, also known as *gycolysis*, is to catabolize glucose to pyruvic acid. The Embden-Meyerhof pathway yields a net gain of two molecules of ATP. The Krebs cycle, also known as the *citric acid cycle*, is a series of biochem-ical reactions that occur in the mitochondria of eukaryotic cells and in the cyto-plasm of prokaryotic cells. The Krebs cycle splits acetyl-CoA into CO_2 and hydrogen atoms. For every molecule of acetyl-CoA that enters the Krebs cycle, a molecule of ATP is produced as well as NADH and FADH which are shuttled to the electron transport chain to make more ATP.

The electron-transport chain, which occurs in the mitochondria of eukary-otic cells and in the cytoplasm of prokaryotic cells, is composed of a series of electron carriers that transfer electrons from donor molecules to a final electron acceptor. The electrons move down an energy gradient. The difference in free energy that occurs between the donor molecules and the acceptor electron releases large amounts of energy. The energy changes that occur eventually produce large amounts of ATP.

Other molecules such as lipids and proteins can also be used to produce ATP.

Lipids consist of glycerol and fatty acids and can produce ATP. Enzymes called *lipases* hydrolyze the bonds attaching the glycerol to the fatty acid chains. Glycerol is converted to DHAP, which is oxidized to pyruvic acid in glycolysis.

The fatty acids are broken down in catabolic reactions called *beta-oxidation*. In beta-oxidation, enzymes split off pairs of hydrogenated carbon atoms that make up the fatty acids. The enzymes then join these pairs to CoA to form acetyl-CoA. Acetyl-CoA is used in the Krebs cycle to generate ATP.

Proteins are split into amino acids. The amino functional group is removed in a reaction called *deamination*. The remaining molecules can then enter into the Krebs cycle.

The Krebs Cycle

The Krebs cycle (Fig. 5-2), also known as the *citric acid cycle* and the *tricarboxylic acid* (TCA) *cycle*, is named for Sir Hans Krebs, a biochemist, who in the 1940s explained how these reactions work. In the Krebs cycle, acetyl-CoA is split into CO_2 and hydrogen atoms. Acetyl CoA contains large amounts of potential energy. Through a series of oxidation and reduction reactions, this

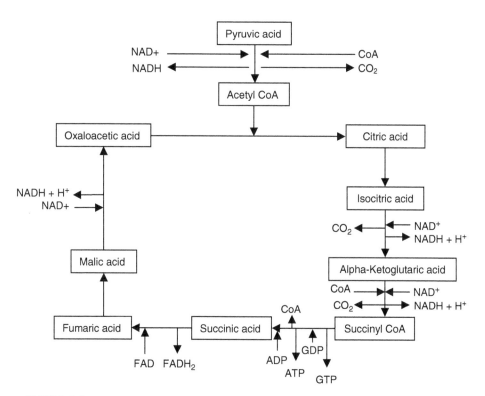

FIGURE 5-2 • The Krebs cycle (citric acid cycle).

potential energy is transferred from electrons to electron carrier coenzymes, mainly the coenzyme NAD+. The coenzymes are reduced (NAD+ is reduced to NADH) and the substances derived from pyruvic acid (the acetyl group) are oxidized.

The first step in the Krebs cycle is the creation of citric acid. This occurs as acetyl CoA enters the Krebs cycle. The CoA will leave the acetyl group, and the acetyl group will combine with oxaloacetic acid. The attachment of the two-carbon acetyl group to the four-carbon oxaloacetic acid group forms the six-carbon citric acid.

The two general types of reactions that take place in the Krebs cycle are decarboxylation and oxidation-reduction reactions. From the breakdown of one six-carbon glucose molecule (glycolysis), the two acetyl CoA molecules that are produced and enter the Krebs cycle will result in six molecules of NADH and two molecules of $FADH_2$ by oxidation-reduction reactions and six molecules of CO_2 from decarboxylation. The CO_2 produced will be lost in the environment as a gas. The NADH and $FADH_2$, which contain the majority of the energy stored in the original glucose molecule, will transfer this energy to ATP. This will occur through a series of reactions called the electron transport chain.

The New Chain Gang: The Electron-Transport Chain

Glycolysis, the intermediate reaction, and the Krebs cycle result in the synthesis of only four ATP molecules. Most of the ATP that is generated comes from the oxidation of NADH and $FADH_2$ in the electron-transport chain.

The electron-transport chain, which occurs in the inner membrane of the mitochondria in eukaryotic cells and in the cytoplasm of prokaryotic cells, is composed of a series of electron carriers that transfer electrons from donor molecules such as NADH and $FADH_2$ to an acceptor atom such as oxygen (O_2). The electrons move down an energy gradient like water flowing down a series of waterfalls in rapids.

The energy changes that occur at several points in the chain are very large and can provide the eventual production of large amounts of ATP. The free energy that electrons have entering the electron-transport chain is greater in the beginning than at the end. It is this energy that enables the protons (H^+) to be pumped across the inner mitochondrial membrane and the inner membrane space.

When the electrons move down the chain, they transfer their energy to the H^+ pumps within the plasma membrane. The reactions of the electron transport chain take place in the inner membrane of the mitochondria in eukaryotic cells or in the plasma membrane in prokaryotic cells. In the mitochondria, this system is set up into four complexes of carriers.

Each of these carriers transports electrons part of the way to O_2 (which is the final electron acceptor). The carriers, coenzyme Q or ubiquinones, cytochromes, and flavoproteins connect these complexes. This process by which energy comes from the electron-transport chain is provided by protons (H^+) and is used to make ATP.

Three ATP molecules can be synthesized from ADP and inorganic phosphate (P_i) when two electrons pass from NADH ultimately to an atom of O_2.

The electron-transport chain uses flavoproteins, cytochromes, and ubiquinone molecules as its carriers. Prokaryotic and eukaryotic electrons work using the same fundamental principles, although they differ in construction.

The electron-transport chain in *Escherichia coli* bacteria, for example, transports electrons from NADH to acceptors and moves protons across the plasma membrane. The *E. coli* electron transport chain is branched and contains different cytochromes. The two branches are cytochrome d and cytochrome o. Coenzyme Q donates electrons to both branches. These chains operate in different conditions. For example, the cytochrome d branch will function when O_2 levels are low and does not actively pump protons, whereas the cytochrome o branch operates in higher O_2 concentrations and is a proton pump.

To summarize, prokaryotic organisms can produce 38 ATP molecules from the complete oxidation of glucose. Eukaryotic organisms, on the other hand, produce 36 molecules from the complete oxidation of glucose.

Fermentation

Fermentation is the partial oxidation of glucose or another organic compound to release energy; it uses an organic molecule as the final electron acceptor. In a simple fermentation reaction, NADH reduces pyruvic acid from glycolysis to form lactic acid. Another example involves two reactions. The first is a decarboxylation reaction where CO_2 is given off (this is the CO_2 that causes bakery items to rise), followed by a subsequent reduction reaction that produces ethanol.

The essential function of fermentation is the regeneration of NAD^+ for glycolysis so that ADP molecules can be phosphorylated to ATP. The benefit of fermentation is that it allows ATP production to continue in the absence of O_2. Microorganisms that ferment can grow and colonize in an anaerobic environment.

Microorganisms produce a variety of fermentation products. The products of fermentation of cells are waste products of the cells, but many are useful to humans. These include *ethanol* (the alcohol that humans can drink, as in beer, wine, and liquor), *acetic acid* (vinegar), and *lactic acid* (found in cheese, sauerkraut, and pickles).

Photosynthesis

Some organisms use anabolic pathways to synthesize organic molecules from inorganic CO_2. Most of these phototrophic organisms are autotrophic and are capable of surviving and growing on carbon dioxide as their only carbon source. The energy from sunlight is used to reduce CO_2 to carbohydrates. This process is called *photosynthesis*. The ability of an organism to photosynthesize depends on the presence of light-sensitive pigments called *chlorophyll* or related compounds. These pigments are found in plants, algae, and certain bacteria.

Photosynthetic organisms can generate their own energy by two separate types of reactions. In *light reactions*, light energy is converted into chemical energy, and in *dark reactions* (light independent reactions), the chemical energy from the light reactions is used to reduce CO_2 to carbohydrates. For the growth and survival of autotrophic organisms, energy is supplied in the form of an ATP molecule. Electrons for the reduction of CO_2 come from nicotinamide adenine dinucleotide phosphate (NADPH). NADPH is produced from the reduction of $NADP^+$ by electrons from electron donors such as water.

Purple and green bacteria use light most of the time to form ATP. Green plants, algae, and cyanobacteria do not use reduced sulfur compounds or organic compounds to obtain reducing power. Instead, they obtain electrons for $NADP^+$ reduction by splitting water molecules. By splitting water, O_2 is produced as a by-product. The reduction of $NADP^+$ to NADPH by these organisms depends on light and therefore is a light-mediated event. Owing to the production of molecular O_2, the process of photosynthesis in these organisms is called *oxygenic photosynthesis*. In contrast, the purple and green bacteria do not produce oxygen. This process is called *anoxygenic photosynthesis*.

Photosynthetic organisms capture light with pigment molecules. An important pigment molecule is chlorophyll. There are different structures of chlorophyll, the most common of which are *chlorophyll a* and *chlorophyll b*. Chlorophyll a is the principal chlorophyll of higher plants, most algae, and cyanobacteria. Purple and green bacteria have chlorophylls of a different structure, called *bacteriochlorophyll*.

Accessory pigments, such as *carotenoids* and *phycobilins*, are also involved in capturing light energy. Carotenoids play a photoprotective role, preventing photooxidative damage to the phototrophic cell. Phycobilins serve as light-harvesting pigments.

Cells arrange numerous molecules of chlorophyll and accessory pigments within membrane systems called *photosynthetic membranes*. The location of these membranes differs between eukaryotic and prokaryotic microorganisms. In eukaryotic organisms, photosynthesis occurs in specialized organelles called *chloroplasts*. The chlorophyll pigments are attached to sheetlike membrane structures of the chloroplasts called *thylakoids*. These thylakoids are arranged in stacks called *grana*.

In prokaryotic organisms, there are no chloroplasts. The photosynthetic pigments are integrated in a membrane system that arises from the cytoplasmic membranes.

Photosystems I and II

Electron flow in oxygenic phototrophs involves two sets of photochemical reactions. Oxygenic phototrophs use light energy to generate ATP and NADPH. The electrons from NADPH come from the splitting of H_2O to get O_2 and electrons (e^-). The two systems of light reactions are called *photosystem I* and *photosystem II*. Photosystems I and II function together in the oxygenic process. Under certain conditions, many algae and some cyanobacteria can carry out cyclic photophosphorylation using only photosystem I. These organisms can obtain reducing power from sources other than water. This requires the presence of anaerobic conditions and a reducing substance such as hydrogen (H_2) or hydrogen sulfide (H_2S). Photosystem II is responsible for splitting water to yield $H_2 + O$.

Light Independent Reactions

Many photosynthetic bacteria contain carboxysomes in their cytoplasm. These carboxysomes contain many copies of the complex enzyme *rubisco* (the most abundant and probably the most important enzyme on the planet) to start the Calvin-Benson cycle (also called the Calvin cycle).

The fixation of CO_2 by most photosynthetic and autotrophic organisms involves the biochemical pathway called the *Calvin cycle*. The Calvin cycle is a reductive, energy-demanding process which uses the energy from ATP to reduce CO_2 to small carbohydrates that are metabolized to glucose and ultimately to more complex carbohydrates such as starch, sucrose, and glycogen. Other substances containing carbon can be used, but CO_2 is preferred.

QUIZ

Fill in the blanks:

1. NAD$^+$ stands for _____.

2. FAD$^+$ stands for _____.

3. NADP$^+$ stands for _____.

4. ATP stands for _____.

5. ADP stands for _____.

Match the reaction:

    A. Photophosphorylation
    B. Hydrolases
    C. Isomerases
    D. Lyases
    E. Transferases

6. Catabolic reaction using water. _____

7. Split molecules without using water in a catabolic reaction. _____

8. Transfer functional groups from one molecule or another substrate in an anabolic reaction. _____

9. Rearrange atoms within a substrate rather than adding or subtracting anything from the reaction. _____

10. Uses energy from sunlight to phosphorylate ADP with organic phosphate. _____

chapter 6

Microbial Growth and Controlling Microbial Growth

In this chapter, you'll learn about how microorganisms grow and how disrupting their growth is used to treat diseases.

CHAPTER OBJECTIVES

In this chapter, you will

- Learn about microbial growth
- Become familiar with culture media
- Examine anaerobic growth
- Learn the phases of growth

You and microorganisms are alike. You both need a variety of elements to form the molecules of life. Without them, you'll die, and so will a microorganism. The chemicals that we're speaking about are called *nutrients*. There are many nutrients needed by you and microorganisms, but the most important nutrients are carbon, hydrogen, nitrogen, and oxygen. These elements make up the organic molecules that make up the proteins, lipids, carbohydrates, and nucleic acids found in cells.

Nutrients are obtained from the environment and from other living things. Microorganisms acquire nutrients by living on or in another organism and using the organism's cells for its nutrients. We commonly refer to this as an infection. The microorganism interferes with your metabolism and disrupts your homeostasis. In other words, you feel sick.

Organisms are classified based on how the organism feeds. Organisms that use carbon dioxide (CO_2) as their source of carbon are called *autotrophs* and feed themselves—*auto-* means "self" and *-troph* means "nutrition." Autotrophs make organic compounds from CO_2 and do not feed on organic compounds from other organisms.

Organisms that obtain carbon from organic nutrients such as proteins, carbohydrates, amino acids, and fatty acids are called *heterotrophs*. Heterotrophic organisms acquire or feed on organic compounds from other organisms. You are a heterotrophic organism because you eat other living things to acquire nutrients for life.

Another way organisms are classified is in terms of whether they use chemicals or light as a source of energy. Organisms that acquire energy from redox reactions involving inorganic and organic chemicals are called *chemotrophs*. Organisms that use light as their energy source are called *phototrophs*.

Chemical Requirements for Microbial Growth

Microorganisms require chemical elements to grow. These elements are carbon, oxygen, nitrogen, hydrogen, phosphorus, and sulfur. These elements are needed to make proteins, lipids, and nucleic acids. Without these elements no growth will occur.

Carbon

Carbon is one of the most important requirements for microbial growth. Carbon is the backbone of living matter.

Oxygen

Microorganisms that use oxygen produce more energy from nutrients than microorganisms that do not use oxygen. Those organisms that require oxygen are called *obligate aerobes*. Oxygen is essential for obligate aerobes because it serves as the final electron acceptor in the electron-transport chain, which produces most of the adenosine triphosphate (ATP) in these organisms. An example of an obligate aerobe is *Micrococcus*. Some organisms can use oxygen when it is present but can continue to grow by using fermentation or anaerobic respiration when oxygen is not available. These organisms are called *facultative anaerobes*. An example of a facultative anaerobe is *Escherichia coli*, which is found in the large intestine of vertebrates, such as humans.

Some bacteria cannot use molecular oxygen and even can be harmed by it. Examples include *Clostridium botulinum*, the bacterium that causes botulism, and *Clostridium tetani*, the bacterium that causes tetanus. These organisms are called *obligate anaerobes*. Molecular oxygen (O_2) is a poisonous gas to obligate anaerobes. Toxic forms of oxygen include

- *Singlet oxygen (O)*. This is an extremely active, high-energy state and is present in cells that use phagocytosis to ingest foreign bacteria cells.
- *Superoxide free radicals (O_2^-)*. These are formed during normal respiration of organisms that use oxygen as a final electron acceptor. In obligate anaerobes, some superoxide free radicals are formed in the presence of oxygen. These superoxide free radicals are incredibly toxic to cellular components. In order for organisms to grow in atmospheric oxygen, they must produce *superoxide dimutase* (SOD) enzymes. Superoxide dimutase neutralizes superoxide free radicals. Aerobic bacteria, facultative anaerobes growing aerobically, and aerotolerant anaerobes produce SOD, which converts the superoxide free radical into molecular oxygen (O_2) and hydrogen peroxide (H_2O_2). This can be seen when you place hydrogen peroxide on a wound infected with bacteria. When hydrogen peroxide is placed on the colony of bacterial cells that produce catalase, oxygen bubbles are released. This is the foaming you see when you place hydrogen peroxide on a cut. Human cells also produce catalase, which converts hydrogen peroxide to water and oxygen.
- *Hydroxyl radical (OH–)*. Also known as a hydroxide ion, this is made in the cytoplasm of the cell by ionizing radiation and by aerobic respiration. Most aerobic respiration produces some hydroxyl radicals. Hydroxyl radical is a toxic form of oxygen.

Culture Media

A *culture medium* is nutrient material prepared in the laboratory for the growth of microorganisms. Microorganisms that grow in size and number on a culture medium are referred to as a *culture*.

In order to use a culture medium, it must be *sterile*, meaning that it contains no living organisms. This is important because we want only microorganisms that we add to grow and reproduce, not others. We must have the proper nutrients, pH, moisture, and oxygen levels (or no oxygen) for a specific microorganism to grow.

Many types of culture media are available for microbial growth. Media are constantly being developed for the identification and isolation of bacteria in the research of food, water, and microbiology studies.

The most popular and widely used medium in microbiology laboratories is the solidifying agent *agar*. Agar is a complex polysaccharide derived from red algae. Very few microorganisms can degrade agar, so it usually remains in a solid form. Agar media usually are contained in test tubes or Petri dishes. The test tubes are held at a slant, and the agar is allowed to solidify on an angle, called a *slant*. A slant increases the surface area for organism growth. When agar solidifies in a vertical tube, it is called a *deep*. The shallow dishes with lids to prevent contamination are called *Petri dishes*. Petri dishes are named after their inventor, Julius Petri, who in 1887 first poured agar into glass dishes.

Chemically Defined Media

For a medium to support microbial growth, it must provide an energy source, as well as carbon, nitrogen, sulfur, phosphorous, and any other organic growth factors (such as amino acids, purines, pyrimidine, and vitamins), for the organism to use.

A *chemically defined* medium is one whose exact chemical composition is known. Chemically defined media *must* contain growth factors required for metabolism. Chemically defined media are used for the growth of autotrophic bacteria and some heterotrophs. Heterotrophic bacteria and fungi normally are grown on *complex media*, which are made up of nutrients such as yeasts, meat extract, plants, or proteins (the exact composition is not quite known and can vary with each mixture). In complex media, the energy, carbon, nitrogen, and sulfur needed for microbial growth are provided by protein. Partial digestion by acids and enzymes breaks down proteins into smaller amino acids called *peptones*.

Peptones are soluble products of protein hydrolysis. These small peptones can be digested by bacteria. Different vitamins and organic growth factors can be provided by meat and yeast extracts.

If a complex medium is in a liquid form, it is called a *nutrient broth*. If agar is added, it is called a *nutrient agar*. Agar is not a nutrient; it is a solidifying agent that enhances colony function.

Anaerobic Growth

Because anaerobic organisms can be killed when exposed to oxygen, they must grow in an environment devoid of O_2. Special media called *reducing media* are often used. Reducing media contain ingredients such as sodium thioglycollate that attach to dissolved oxygen and deplete the oxygen in the culture medium. Obligate anaerobes can also be grown in anaerobic chambers.

Selective and Differential Media

In health clinics and hospitals, it is necessary to identify microorganisms that are associated with disease. Selective and differential media therefore are used. *Selective media* are made to encourage the growth of some bacteria while inhibiting others. An example of this is bismuth sulfite agar. Bismuth sulfite agar is used to isolate *Salmonella typhi* from fecal matter. *S. typhi* is a gram-negative bacterium that causes salmonellosis, a type of food poisoning. *Differential media* make it easy to distinguish colonies of desired organisms from nondesirable colonies growing on the same plate. Pure cultures of microorganisms have identifiable reactions with different media. An example is blood agar. Blood agar is a dark red/brown medium that contains red blood cells and is used to identify bacterial species that *destroy* red blood cells. An example of this type of bacterium is *Streptococcus pyogenes*, the agent that causes strep throat.

MacConkey agar is both selective and differential. MacConkey agar contains bile salts and crystal violet, which inhibit the growth of gram-positive bacteria, and lactose, in which some gram-negative bacteria can grow.

Enrichment cultures are usually liquids and provide nutrients and environmental conditions that provide for the growth of certain microorganisms but not others.

Pure Cultures

Infectious material or materials that contain pathogenic microorganisms include pus, sputum, urine, feces, soil, water, and food. These infectious materials can

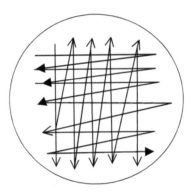

FIGURE 6-1 • Streak-plate method used to isolate bacteria.

contain numerous kinds if bacteria. If a mixed population, like that found in fecal matter, is placed on solid medium, individual bacterium will divide and form a colony. Each colony is made up of thousands of bacterium, all derived from a single bacterial cell. Each cell in a bacterial colony is genetically identical. *Microbial colonies* have distinct appearances that distinguish one microorganism from another.

The *streak-plate method* is the most common way to get pure cultures from a mixed population. A device called an *inoculating loop* is sterilized and dipped into a culture of a microorganism or microorganisms and then is streaked in a pattern over a nutrient medium. As the pattern is made, bacteria are rubbed off from the loop onto the nutrient medium. The last cells that are rubbed off the loop onto the medium are far enough apart to allow isolation of separate colonies of the original culture (Fig. 6-1).

Preserving Bacterial Cultures

Two common methods of preserving microbial cultures are *deep freezing* and *lyophilization* (or *freeze drying*). Deep freezing is a process in which a pure culture of microorganisms is placed in a suspending liquid and frozen quickly at temperatures ranging from –50 to –95°C. With this type of freezing method, cultures usually can be thawed and used even after several years. Lyophilization, or freeze drying, quick freezes suspended microorganisms at temperatures from –54 to –72°C while water is removed by using a high-pressure vacuum. While under the vacuum, the container is sealed with a torch. The surrounding microbes in the sealed container can last for years. The organisms can be retained and revived by hydrating them and placing them into a liquid nutrient medium.

Growing Bacterial Cultures

Bacteria normally reproduce by a process called *binary fission. In general the steps of binary fission are as follows:*

1. The cell elongates, and chromosomal DNA is replicated.
2. The cell wall and cell membrane pinch inward and begin to divide.
3. The pinched parts of the cell wall meet, forming a cross-wall completely around the divided DNA.
4. The cells separate into two individual cells.

Some bacteria reproduce by *budding.* A small outgrowth or bud emerges from the bacterium and enlarges until it reaches the size of the daughter cell. It then separates, forming two identical cells. Some bacteria, called *filamentous bacteria* (or *actinomycetes*), reproduce by producing chains of spores located at the tips of the filaments. The filaments fragment, and the fragments initiate the growth of new cells.

Generation Time

The *generation time* is the amount of time needed for a cell to divide. This varies among organisms and depends on the environment they are in and the temperature of the environment. Some bacteria have a generation time of 24 hours, although the generation time of most bacteria is between one and three hours. Bacterial cells grow at a rapid rate. For example, with binary fission, certain bacteria such as *E. coli* can double every 20 minutes. In 30 generations of bacteria (10 hours), the number could reach 1 billion. It is difficult to graph population changes of this magnitude using arithmetic numbers, so logarithmic scales are used to graph bacterial growth.

Phases of Growth

There are four basic phases of bacterial growth: the lag phase, the log phase, the stationary phase, and the death phase.

The Lag Phase

In the *lag phase*, there is little or no cell division. This phase can last from one hour to several days after bacteria are introduced to a nutrient-rich environment. Here, the microbial population is involved in intense metabolic activity involving DNA and enzyme synthesis. This is like a factory shutting down for two weeks in the summer for renovations. New equipment is replacing old and employees are working, but no product is being turned out.

The Log Phase

In the *log phase*, cells begin to divide and enter a period of growth or logarithmic increase. This is the time when cells are the most active metabolically. This is the time when the product of the factory must be produced in an efficient matter. In this phase, however, microorganisms are very sensitive to adverse conditions of their environment.

The Stationary Phase

This phase is one of equilibrium. The growth rate slows, the number of dead microorganisms equals the number of new microorganisms, and the population stabilizes. The metabolic activities of individual cells that survive will slow down. The reasons why the growth of the organisms stops is possibly that the nutrients have been used up, waste products have accumulated, and drastic harmful changes in the pH of the organism's environment have occurred.

There is a device called a *chemostat* that drains off old, used medium and adds fresh medium. In this way, a population can be kept in the growth phase indefinitely.

The Death Phase

Here, the number of dead cells exceeds the number of new cells. This phase continues until the population is diminished or dies out entirely.

Still Struggling

Sometimes the phases of growth may be difficult to understand. Remember there are four phases:

1. *Lag phase.* Little or no cell division occurs, but cells have high metabolic activity.
2. *Log phase.* Cell division begins, and the cell enters a period of exponential growth.

3. *Stationary phase.* The growth rate slows, and the population stabilizes. The number of dead and new organisms is equal.

4. *Death phase.* The number of dead cells becomes greater then the number of new cells. The population starts to diminish.

Measurements of Microbial Growth

A *plate count* is the most common method of determining the number of living cells. This method measures the number of viable cells. Plate counts may take 24 hours or more for visible colonies to form.

A plate count is performed by either a pour-plate method or a spread-plate method. With the pour-plate method, either 1.0 or 0.1 milliliter of a bacterial solution is placed into a Petri dish. Melted nutrient agar is added, which then is agitated gently or mixed. When the agar has solidified, the plate is then incubated. With this technique, heat-sensitive microorganisms can be damaged by the melted agar and be unable to form colonies. To avoid death of cells owing to heat, the spread-plate method is used mostly. Here, a 0.1-milliliter bacterial solution is added to the surface of a prepoured, solid nutrient agar. The bacterial solution then is spread evenly over the medium.

When the amount of bacteria is very small, as in lakes and pure streams, bacteria can be counted by *filtration methods.* Here, 100 milliliters of water is passed through a thin membrane, whose pores are too small for the bacteria to pass through. The bacteria that are retained on the filter are placed on a Petri dish containing a pad soaked with liquid nutrient. An example of bacteria that are grown using this method is *coliform bacteria,* which are indicators of fecal pollution of food or water.

When using the *direct microscopic count method,* a measured volume of bacteria is suspended in a liquid placed inside a designated area of a microscopic slide. For example, a 0.01-milliliter sample is spread over a marked square centimeter of a slide, stained, and viewed under the 100× objective lens. The area for the viewing field is obtained. Once the number of bacteria is obtained or determined in several different fields, an average can be taken of the number of bacteria per viewing field. From these data, the number of bacteria in the square centimeter over which the sample has been spread can be calculated. Because the area on the slide contained 0.01 milliliter of sample, the number of bacteria in each milliliter of the suspension is the number of bacteria in the sample times 100.

Establishing Bacterial Numbers by Indirect Methods

Not all microbial cells must be counted to establish their number. With some types of work, estimating the turbidity is a practical way of monitoring microbial growth. *Turbidity* is the cloudiness of a liquid or the loss of transparency because of insoluble matter.

The instrument used to measure turbidity is the *spectrophotometer* or *colorimeter*. In the spectrophotometer, a beam of light is transmitted through the bacteria that are suspended in the liquid medium to a photoelectric cell. As bacteria growth increases, less light will reach the photoelectric cell. The change in light will register on the instrument's scale as the percentage of transmission. The amount of light striking the light-sensitive detector on the spectrophotometer is inversely proportional to the number of bacteria under normal cultures: The less light, the more bacteria.

Another indirect way of measuring bacterial growth is to measure the *metabolic activity* of the colony. In this method, it is assumed that metabolic waste products, carbon dioxide (CO_2) and acid, are in direct proportion to the number of bacteria present. The more bacteria, the more waste products present.

For filamentous organisms, such as molds, a way to measure growth is by *dry weight*. The fungus is removed from its growth medium, filtered, and placed in a weighing bottle dried in a desiccator (a *desiccator* is a device that removes water). With bacteria, the culture is removed from the medium by centrifugation.

Controlling Microbial Growth

It is very important to control microbial growth in surgical and hospital settings, as well as in industrial and food-preparation facilities. There are many terms used to describe the fight to control microorganisms.

Sterilization is the destruction of all microorganisms and viruses, as well as endospores. Sterilization is used in preparing culture media and canned foods. It is usually performed by steam under pressure, incineration, or a sterilizing gas such as ethylene oxide.

Commercial sterilization is the treatment to kill endospores in commercially canned products. An example is the bacteria *C. botulinum*, which causes botulism.

Antisepsis is the reduction of pathogenic microorganisms and viruses on living tissue. Treatment is by chemical antimicrobials, such as iodine and alcohol. Antisepsis is used to disinfect living tissues without harming them.

Aseptic means to be free of pathogenic contaminants. Examples include proper hand washing, flame sterilization of equipment, and preparing surgical environments and instruments.

Any word with the suffix *-cide* or *-cidal* indicates the death or destruction of an organism. For example, a *bactercide* kills bacteria. Other examples are fungicides, germicides, and virucides. Germicides include detergents and soaps, hydrogen peroxide, alcohol, heavy metals, aldehydes, and gases such as ethylene oxide and propylene oxide.

Disinfection is the destruction or killing of microorganisms and viruses on nonliving tissue by the use of chemical or physical agents. Examples of these chemical agents are phenols, alcohols, aldehydes, and surfactants.

Degerming is the removal of microorganisms by mechanical means, such as cleaning the site of an injection. This area of the skin is degermed by using an alcohol wipe or a piece of cotton swab soaked with alcohol. Hand washing also removes microorganisms by chemical means.

Pasteurization, as noted in Chapter 1, uses heat to kill pathogens and reduce the number of food spoilage microorganisms in foods and beverages. Examples are pasteurized milk and juice.

Sanitization is the treatment to remove or lower microbial counts on objects such as eating and drinking utensils to meet public health standards. This is usually accomplished by washing the utensils in high-temperature or scalding water and disinfectant baths. Bacteriostatic, fungistatic, and virustatic agents—or any word with the suffix *-static* or *-stasis*—indicate the inhibition of a particular type of microbial growth. These are unlike bactericides or fungicides that kill or destroy the organism. Germistatic agents include refrigeration, freezing, and some chemicals.

Microbial Death Rates

Microbial death is the term used to describe the permanent loss of a microorganism's ability to reproduce under normal environmental conditions. A technique for the evaluation of an antimicrobial agent is to calculate the *microbial death rate*. When populations of a particular organism are treated with heat or antimicrobial chemicals, they usually die at a constant rate.

The effectiveness of antimicrobial treatments is influenced by the number of microbes that are present. The larger the population, the longer it takes to destroy it.

Organic matter, such as blood, saliva, or fecal matter, inhibit the action of chemical antimicrobials. Time of exposure to heat or radiation is also important. Many chemical antimicrobials need longer exposure times to be effective in the death of more resistant microorganisms or endospores.

Action of Antimicrobial Agents

There are two categories into which chemical and physical antimicrobial agents fall: those which affect the cell walls or cytoplasmic membranes of the microorganism and those which affect cellular metabolism and reproduction. As stated in Chapter 4, the cell wall is located outside the microorganism's plasma membrane. The cell's plasma membrane regulates substances that enter and exit the cell during its life. Nutrients enter the cell as waste products exit the cell. Damage to the plasma membrane proteins or phospholipids by physical or chemical agents allows the contents of the cell to leak out. This causes the death of the cell.

Proteins act as regulators in cellular metabolism, function as enzymes (which are important in all cellular activities), and form structural components in cell membranes and cytoplasm. The function of a protein depends on its three-dimensional shape. The hydrogen and disulfide bonds between the amino acids that make up the protein maintain this shape. Extreme heat, certain chemicals, and very high or low pH easily can break some of these hydrogen bonds. This breakage is referred to as the *denaturing* of the protein. The protein's shape is changed, thus affecting the function of the protein and ultimately bringing death to the cell.

Certain chemicals, radiation, and heat can damage nucleic acids. The nuclear acids, DNA and RNA, carry the cell's genetic information. If these are damaged, the cell can no longer replicate or synthesize protein enzymes, which are important in cell metabolism.

Chemical Agents that Control Microbial Growth

The growth of a microorganism can be controlled through the use of a chemical agent. A *chemical agent* is a chemical that either inhibits or enhances the growth of a microorganism. Commonly used chemical agents include phenols, phenolics, glutaraldehyde, and formaldehyde.

Phenols and Phenolics

Phenols are compounds derived from phenol (carbolic acid) molecules. Pheno-lics disrupt the plasma membrane by denaturing proteins; they also disrupt the plasma membrane of the cell. As mentioned in Chapter 1, Joseph Lister used phenol in the late 1800s to reduce infection during surgery.

Alcohols are effective against bacteria, fungi, and viruses. However, they are not effective against fungal spores or bacterial endospores. Alcohols that are used commonly are isopropanol (rubbing alcohol) and ethanol (the alcohol we drink).

Alcohols denature proteins and disrupt cytoplasmic membranes. Pure alcohol is not as effective as 70% alcohol because the denaturation of proteins requires water. Alcohols are good to use because they evaporate rapidly. A disadvantage is that they may not contact the microorganisms long enough to be effective. Alcohol is used commonly in swabbing the skin prior to an injection.

Halogens are nonmetallic, highly resistive chemical elements. Halogens are effective against vegetative bacterial cells, fungal cells, fungal spores, protozoan cysts, and many viruses. Halogen-containing antimicrobial agents include iodine, which inhibits protein function. Iodine is used in surgery and by campers to disinfect water. An *iodophor* is an iodine-containing compound that is longer lasting than iodine and does not stain the skin. Other halogen agents include

- *Chlorine (Cl_2).* Used to treat drinking water and swimming pools and in sewage plants to treat waste water. Chlorine products such as sodium hypochlorite (household bleach) are effective disinfectants.
- *Chlorine dioxide (ClO_2).* A gas that can disinfect large areas.
- *Chloroamines.* Chemicals containing chlorine and ammonia. They are used as skin antiseptics and in water supplies.
- *Bromine.* Used to disinfect hot tubs because it does not evaporate as quickly as chlorine at high temperatures.
- *Oxidizing agents.* Kill microorganisms by oxidizing their enzymes, thus preventing metabolism. Hydrogen peroxide, for example, disinfects and sterilizes inanimate objects, such as food-processing and medical equip-ment, and it is also used in water purification.

Arsenic, zinc, mercury, silver, nickel, and copper are *heavy metals* owing to their high molecular weights. They inhibit microbial growth because they dena-ture enzymes and alter the three-dimensional shapes of proteins. Heavy metals are bacteriostatic and fungistatic agents.

An example is silver nitrate. At one time, hospitals required newborn babies to receive a 1% cream of silver nitrate to their eyes to prevent blindness caused by *Neisseria gonorrhoeae*, which could enter the baby's eyes while passing through the birth canal of a mother who was infected. Today, antibiotic ointments that are less irritating are used. Another example is the use of copper in swimming pools, fish tanks, and reservoirs to control algae growth. Copper interferes with chlorophyll, thus affecting metabolism and energy.

Glutaraldehyde and Formaldehyde

Aldehydes function in microbial growth by denaturing proteins and inactivating nucleic acids. Two types are glutaraldehyde, which is a liquid, and formaldehyde, which is a gas.

Glutaraldehyde is used in a 2% solution to kill bacteria, fungi, and viruses on medical and dental equipment. Health care workers and morticians dissolve gaseous formaldehyde in water, making a 37% solution of *formalin*. Formalin is used in disinfecting dialysis machines and surgical equipment and in embalming bodies after death.

Gaseous agents, such as ethylene oxide, propylene oxide, and beta-propiolactone, are used on equipment that cannot be sterilized easily with heat, chemicals, or radiation. Certain items, such as pillows, mattresses, dried or powdered food, plastic-ware, sutures, and heart-lung machines, are placed in a closed chamber that is then exposed to these gases. Gaseous agents denature proteins.

Surfactants

Surfactants are chemicals that act on surfaces by decreasing the tension of water and disrupting cell membranes. Examples are household soaps and detergents.

QUIZ

How do we control microbial growth?

1. This process is used to prepare cultured media and canned foods, performed by steam under pressure or sterilizing gas. _____

2. This process is used to disinfect living tissues without harming them by using chemical antimicrobials. _____

3. This treatment is used to kill endospores in commercially canned products. _____

4. The destruction of microorganisms on nonliving tissue by the use of chemical and physical agents is called _____.

5. The removal of microorganisms by mechanical means is called _____.

Match the chemical agent:

    A. Chlorine
    B. Chlorine dioxide
    C. Hydrogen peroxide
    D. Bromine
    E. Oxidizing agents

6. Used to kill microorganisms by oxidizing their enzymes. _____

7. Used as a disinfectant and in the treatment of drinking water and pools. _____

8. Used to disinfect large areas. _____

9. Used to disinfect hot tubs. _____

10. Used as an antiseptic for skin. _____

chapter 7

Microbial Genetics

In this chapter, you'll learn about how microorganisms pass along instructions to the next generation, including instructions on how to avoid medical treatment designed to kill the microorganism.

CHAPTER OBJECTIVES

In this chapter, you will

- Be introduced to genetics
- Learn about DNA/RNA
- Examine the control of genes
- Become familiar with the operon model

No doubt you remember the TV commercial where they said your high cholesterol could be caused by your cheeseburger diet—or caused by Uncle Bob. This is not to say that Uncle Bob made the cheeseburgers. The point is that high cholesterol can be caused by genetics. *Genetics* is the study of how characteristics are passed down through generations.

Although we tend to think of genetics in terms of human traits, every living thing receives genetic information passed down from previous generations. *Genetic information* stored in the DNA is similar to a computer program, providing the next generation with instructions on how to stay alive and reproduce the next generation.

However, unlike a computer program, genetics passes along some traits and not others. That is to say, some instructions are not given to the next generation possibly because those instructions are no longer valid owing to the fact that the species has learned to survive without those instructions.

Microorganisms pass along genetic traits to new generations of their species. Those traits enable the species to survive as the environment changes and to know how to process food, excrete waste, and reproduce.

Genetics

Genetics is the branch of science that studies heredity and how traits (expressed characteristics) are passed to new generations of species and between microorganisms. Scientists who study genetics are called *geneticists* and are interested in how traits are expressed within a cell and how traits determine the characteristics of an organism.

Think of a trait as a set of instructions that tells an organism how to do something, such as how to form a toe. Each instruction is contained in a *gene*. As you can imagine, thousands of genes (instructions) are necessary for an organism to grow and flourish. This is why if a youngster looks and behaves like her mother, family members tend to say she has her mother's genes—that is, she has more genes (instructions) from her mom than from her dad.

Genes actually are made up of segments or sections of deoxyribonucleic acid (DNA) or, in the case of a virus, ribonucleic acid (RNA) molecules. Viruses can have RNA. These segments are placed in a specific sequence that codes for a functional product.

DNA Replication: Take My Genes, Please!

In 1868, Swiss biologist Friedrich Miescher carried out chemical studies on the nuclei of white blood cells in pus. His studies led to the discovery of DNA. DNA was not linked to hereditary information until 1943 when work performed by Oswald Avery, Colin MacLeod, and Maclyn McCarty at the Rockefeller Institute revealed that DNA contained genetic information. These studies also revealed that genetic information is passed from *parent cells* to *daughter cells*, creating a pathway through which genetic information is passed to the next generation of an organism.

Scientists were baffled about how the exchange of DNA occurred. The answer came in 1953 when American geneticist James Watson and English physicist Francis Crick discovered the double-helical structure of DNA at the University of Cambridge in England. Discovery of the double-helical structure was the key that enabled Watson and Crick to learn how DNA is replicated.

In the late 1950s, Mathew Meselson and Franklin Stahl first described the DNA molecule and how DNA replicates in a process called *semiconservative replication*. DNA is replicated by taking the parent's double-stranded DNA molecule, unzipping it, and building two identical daughter molecules using the two parental strands as templates. Bases along the two strands of double-helical DNA complement each other. One strand of the pair acts as a template for the other.

DNA replication requires complex cellular proteins that direct the sequence of replication. Replication begins when the parent double-stranded DNA molecule unwinds; then the two strands separate. The DNA polymerase enzyme uses a strand as a template to make a new strand of DNA. The DNA polymerase enzyme proofreads the new DNA and removes bases that do not match and then continues DNA synthesis.

The point at which the double-stranded DNA molecule unzips is called the *replication fork* (Fig. 7-1). The two new strands of DNA each have a base sequence complementary to the original strand. Each double-stranded DNA molecule contains one original and one new strand. In bacteria, each daughter cell receives a circular chromosome that is identical to the parental chromosome.

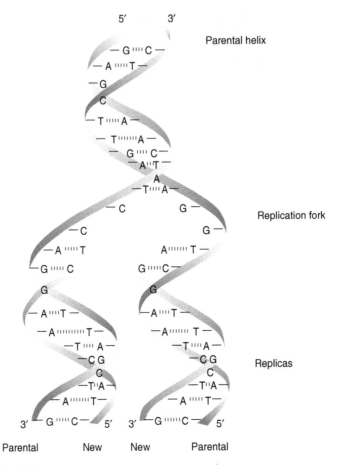

FIGURE 7-1 · In semiconservative replication, new strands are synthesized after the replication fork.

The Chromosome Connection

Chromosomes are structures that contain DNA. *DNA* consists of two long chains of repeating nucleotides that twist around each other, forming a double helix. *Nucleotides* are the building blocks of DNA. A nucleotide in a DNA chain consists of a *nitrogenous base*, a *phosphate group*, and *deoxyribose* (pentose sugar).

The two DNA chains are held together by *hydrogen bonds* between their nitrogenous bases. There are two major types of nitrogenous bases. These are *purines* and *pyrimidines*. There are two types of purine bases: *adenine* (A) and *guanine* (G). There are also two types of pyrimidine bases: *cytosine* (C) and *thymine* (T). Purine and pyrimidine bases are found in both strands of the double helix.

TABLE 7-1 Complementary Messenger RNA Bases and DNA Bases	
Messenger RNA Base	**Double-Helical Strand DNA Molecule Base**
Adenine (A)	Thymine (T)
Guanine (G)	Cytosine (C)
Cytosine (C)	Guanine (G)
Vracil (U)	Adenine (A)

Base pairs are arrangements of nitrogenous bases based on hydrogen bonding. *Adenine* pairs with *thymine*, and *cytosine* pairs with *guanine*. Adenine is said to be complementary to thymine, and cytosine is said to be complementary to guanine. This is known as *complementary base pairing* and is shown in Table 7-1.

Genetic information is encoded by the sequence of bases along a strand of DNA. A nucleotide sequence is ultimately translated into an amino acid, which is the basis of *protein synthesis*. The translation of genetic information from genes to specific proteins occurs in cells.

Protein Synthesis

Protein synthesis is the making of a protein and requires *ribonucleic acid* (RNA), which is synthesized from DNA and uses nucleotides that contain the bases A, C, G, and U. There are three types of RNA. These are

- *Ribosomal RNA (rRNA)*. With other proteins, this makes up part of the ribosomes of the cell.
- *Transfer RNA (tRNA)*. This is needed to transport amino acids to the ribosomes to synthesize proteins.
- *Messenger RNA (mRNA)*. This carries the genetic information from DNA into the cytoplasm to ribosomes, where the amino acids are arranged and proteins are made.

Enzymes called *RNA polymerase* are required to make (synthesize) rRNA, tRNA, and mRNA.

Protein synthesis begins with the *transcription process*, in which DNA sequences are replicated producing mRNA. The mRNA carries genetic information from the DNA to ribosomes. *Ribosomes* are organelles that are the site of protein synthesis.

Nucleotides contained in DNA are duplicated by enzymes before cell division, enabling genetic information to be carried between cells and from one generation to the next. This is referred to as *gene expression* and happens in RNA only.

During transcription, the bases A, C, G, and U pair with complementary bases of the DNA strand that is being transcribed. The G base in the DNA template pairs with the C base in the mRNA. The C base in the DNA template pairs with the G base in the mRNA. The T base in the DNA template pairs with the A base in the mRNA. The A base in the DNA template pairs with the U base in the mRNA. This happens because RNA contains a U base instead of a T base.

Transcription begins when the RNA polymerase binds to DNA at the *promotor site*. The DNA unwinds. One of the DNA strands, called the RNA *coding strand*, serves as a template for RNA synthesis. RNA is synthesized by pairing free nucleotides of the RNA with nucleotide bases on the DNA template strand. The RNA polymerase moves along the DNA as the new RNA strand grows. This continues until the RNA polymerase reaches the terminator site on the DNA or is physically stopped by a section of RNA transcript. The new single-stranded mRNA and the RNA polymerase are released from the DNA.

Here is what's happening: Information found on the mRNA is used by ribosomes to assemble a protein. The nucleotides of the mRNA are read in sequence of three AUG known as codons. The start of mRNA is AUG, the "start" codon. The sequence of codons on the RNA determines the sequence that amino acids are used to synthesize proteins.

Once the transcription process is completed, information on the mRNA is turned into protein in the translation process. The *translation process* is one in which genetic information encoded in mRNA is translated into a specific sequence of amino acids that produce proteins.

Appropriate amino acids are brought to the translation site on the ribosomes and are assembled into a growing chain. It is here that tRNA recognizes specific codons. Each tRNA molecule has an anticodon, which is a sequence of three bases that is complementary to the bases on the codon. These bases then are paired. The tRNA carry specific amino acids to the mRNA. The anticodon allows correct pairing between the information of the mRNA codon and the anticodon of tRNA.

This process continues until a polypeptide is produced. The polypeptide is removed from the ribosome for further processing. The polypeptide may be stored in the Golgi body of a eukaryotic organism. The mRNA molecule degenerates, and the nucleotides are returned to the nucleus. The tRNA molecule is returned to the cytoplasm and combines with new molecules of amino acids.

Still Struggling

Types of ribonucleic acids (RNAs) can be confusing. RNA is important in the making of protein. The three types of RNA are

1. *Ribosomal RNA (rRNA).* This makes up part of the ribosomes of the cell.
2. *Transfer RNA (tRNA).* This transports amino acids from the cytoplasm of the cell to the ribosomes, where they will be utilized for the making of protein.
3. *Messenger RNA (mRNA).* This carries the genetic information from DNA and directs the arrangement of amino acids in making protein.

Genotype and Phenotype: Realizing Your Potential

The genetic makeup of an organism is called a *genotype* and represents that organism's potential properties. Some properties may not develop. Those which do develop are called an organism's *phenotype*. The phenotype represents expressed properties, such as blue eyes and curly hair.

A genotype is the organism's DNA (a collection of genes). The phenotype is based on a collection of proteins. Most of the cell's properties comes from the structures and functional properties of these proteins.

Controlling Genes: You're under My Spell

The process of making proteins (polypeptides become proteins either after they are combined with other polypeptides or when they become biologically functional) begins with copying of the genetic information found in DNA into a complementary strand of RNA. This copying is called *transcription*. mRNA will carry the coded information or instructions for assembling the polypeptides from DNA to the ribosomes where the polypeptides will be made.

The actual building of polypeptides is called *translation*. Translation involves the deciphering of nucleic acid information and converting that information into a language that the proteins can understand.

The Operon Model

In 1961, Francois Jacob and Jacques Monod formulated the *operon model*, which describes how transcription of mRNA is regulated. Transcription of mRNA can be regulated in two ways: repression and induction.

Repression inhibits gene expression and decreases the synthesis of enzymes. Proteins called *repressors* stop the ability of RNA polymerase to initiate transcription from repressed genes. *Induction* activates transcription by producing *inducer*, which is the chemical that induces transcription.

Jacob and Monod identified genes in *Escherichia coli* as *structural genes*, *regulatory genes*, and *control regions*. Collectively, these form a functional unit called the *operon*. Certain carbohydrates can induce the presence of enzymes needed to digest those carbohydrates.

For example, when lactose is present, *E. coli* synthesize enzymes needed to break down lactose. Lactose is an inducer molecule. If lactose is absent, a regulator gene produces a repressor protein that binds to a control region called the *operator site*, preventing the structural genes from encoding the enzyme for lactose digestion. If present, lactose binds to the repressor at the operator site when lactose is present, freeing the operator site. The structural genes are then released and produce their lactose-digesting enzymes.

Mutations: Not a Pretty Copy

A *mutation* is a permanent change in the DNA base sequence (Table 7-2). Some mutations have no expressive effect, whereas other mutations have an expressive effect. When a gene mutates, the enzyme encoded by the gene can become less active or inactive because the sequence of the enzyme amino acids may have changed. The change can be harmful or fatal to the cell, or it can be beneficial—especially if the mutation creates a new metabolic activity.

The most common type of mutation is a *point mutation*, which is also known as *base-substitution mutation*. Point mutations occur when an unexpected base is substituted for a normal base, causing alteration of the genetic code, which then is replicated.

If the mutated gene is used for protein synthesis, the mRNA transcribed from the gene carries the incorrect base for that position. The mRNA may

TABLE 7-2 Types of Mutations	
Type of Mutation	**Description**
Point mutation	Also known as *base substitution*, this is the most common type of mutation and involves a single base pair in the DNA molecule. In a point mutation, a different base is substituted for the original base, causing the genetic code to be altered. The substituted base pair is used when DNA is replicated or transcribed.
Missense mutation	A mutation when a new amino acid is substituted in the final protein by the messenger RNA during transcription.
Nonsense mutation	A mutation when a terminator codon in the messenger RNA appears in the middle of a genetic message instead of at the end of the message, which causes premature termination of transcription.
Frame-shift mutation	Pairs of nucleotides are either added or removed from a DNA molecule.
Loss-of-function mutation	This mutation causes a gene to malfunction.
Spontaneous mutation	Naturally occurring mutation that happens without the presence of a mutation-causing agent.
Induced mutation	Induced in a laboratory.

insert an incorrect amino acid in the protein. If this happens, the mutation is called a *missense mutation*.

Mutations that change or destroy the genetic code are called *nonsense mutations*. If nucleotides are added or deleted from mRNA, the mutation is called a *frame-shift mutation*.

A mutation occurring in the laboratory is called an *induced mutation;* mutations occurring outside the laboratory are called *spontaneous mutations*. An induced mutation occurs when a mutation-causing chemical is present, while a spontaneous mutation occurs for no apparent reason.

Base-substitution and frame-shift mutations occur spontaneously. Agents in the environment or those introduced by industrial processing can directly or indirectly cause mutations. These agents are called *mutagens*. Any chemical or physical agent that reacts with DNA potentially can cause mutations.

Certain mutations make microorganisms resistant to antibiotics or increase their pathogenicity. There are many naturally occurring mutagens, such as radiation from x-rays, gamma rays, and ultraviolet light. These rays break the covalent bonds between certain bases of DNA.

Ultraviolet light binds together adjacent thymines in a DNA strand, forming *thymine dimers* that cannot function in protein synthesis. Unless repaired, these dimers cause damage or death to cells owing to improper transcription or replication of DNA. Some bacteria can repair damage caused by ultraviolet radiation by employing light-repairing enzymes that separate the dimer into the original two thymines. This process is called *photoreactivation*.

Mutation Rate

Mutations occur naturally and can be induced by mutation-causing agents in the environment. However, not all cells experience mutation even if they are exposed to mutation-causing agents.

Scientists measure the impact that mutation has on an organism by determining the mutation rate. The *mutation rate* is the number of mutations per cell division. For example, suppose that you observe the growth of 100 cells that began from a parent cell. If 90 of those cells replicate the parent cell and 10 cells are mutations, then the mutation rate is 10 percent.

Measuring the mutation rate is a way to compare the number of mutations that occur naturally to the number of mutations that occur when a cell is exposed to a potential mutation-causing agent.

First, scientists measure the mutation rate that occurs naturally when a cell is not exposed to a potential mutation-causing agent. Next, the mutation rate is calculated when a cell is exposed to a potential mutation-causing agent. The results of these two observations are compared. If both mutation rates are relatively the same, then the substance being tested is not a mutation-causing agent. However, the substance is a mutation-causing agent if its mutation rate is appreciably higher than the natural mutation rate.

QUIZ

Fill in the blanks:

1. The number of mutations per cell division is called the _____.

2. These two scientists formulated the operon model. _____ and _____

3. The copying of genetic information into a complementary strand of RNA is called _____.

4. The assembly or actual building of polypeptides is called _____.

Match the mutation:

    A. Point mutation
    B. Missense mutation
    C. Nonsense mutation
    D. Frame-shift mutation
    E. Loss-of-function mutation
    F. Spontaneous mutation
    G. Induced mutation

5. This type of mutation is induced in a laboratory. _____

6. This is the most common type of mutation. _____

7. This is a naturally occurring mutation. _____

8. This mutation occurs when a new amino acid is substituted in the final protein by the messenger RNA during transcription. _____

9. This mutation occurs when a terminator codon in the mRNA appears in the middle of the genetic message instead of at the end of the message. _____

10. This type of mutation occurs when pairs of nucleotides are either added or removed from a DNA molecule. _____

11. This type of mutation causes a gene to malfunction. _____

chapter 8

Recombinant DNA Technology

In this chapter, you'll learn about genetic engineering. *Genetic engineering* is the area of microbiology that reorders genetic information by recombining DNA using recombinant DNA technology. Recombinant DNA technology enables scientists to reprogram microorganisms.

CHAPTER OBJECTIVES

In this chapter, you will

- Learn about artificial DNA
- Be introduced to gene therapy
- Examine DNA fingerprinting
- Look at recombinant DNA technology

Deoxyribonucleic acid (DNA) is like a computer program where genes are lines of computer instructions used to tell our bodies how to develop specific characteristics such as how to change food into energy, how to walk, and how to be human. Microorganisms also have DNA that enables them to express specific characteristics that are common to microorganisms. You can say that DNA separates you from looking and behaving like a microorganism.

Genetic information is instructions written by linking together nucleic acids in a specific sequence. Imagine changing the nucleic acid sequence. You would be rewriting instructions on how the organism looks and performs. You could repress undesirable characteristics and encourage expression of desirable characteristics. Scientists change the nucleic acid sequence to change a disease-causing microorganism into a friendly microorganism.

Genetic Engineering: Designer Genes

The modification of an organism's genetic information by changing its nucleic acid genome is called *genetic engineering* and is accomplished by methods known as *recombinant DNA technology*. Recombinant DNA technology opens up totally new areas in research and applied biology and is an important part of biotechnology, a field that is growing. *Biotechnology* is the term used for processes in which organisms are manipulated at the genetic level to form products for medicine, agriculture, and industry.

Recombinant DNA is DNA with a new sequence formed by joining fragments from different sources. One of the first breakthroughs leading to recombinant DNA, or rDNA, technology was the discovery of microbial enzymes that make cuts into the double-stranded DNA. These were discovered by Werner Arber, Hamilton Smith, and Dan Nathans in the late 1960s. These enzymes recognize and cleave specific sequences of four to eight base pairs and are known as *restriction enzymes*. These enzymes recognize specific sequences in DNA and then cut the DNA to produce fragments called *restriction fragments*. The enzymes cut the bonds of the DNA backbone at a point along the exterior of the DNA strands.

There are three types of restriction enzymes. Types I and III cleave DNA away from recognition sites. Type II restriction endonucleases cleave DNA at specific recognition sites. The type II enzymes can be used to prepare DNA fragments containing specific genes or portions of genes. A *gene* can be defined as a segment of DNA (a *segment* is a sequence of nucleotides) that codes for a functional product.

EcoRI cleaves the DNA between guanine (G) and adenine (A) in the base sequence GAATTC. In the double-stranded condition, the base sequence

Enzyme	Microbial Source	Recognition Sequence	Cleavage Sites ($\downarrow,\uparrow$)	End Product
EcoRI	*Escherichia coli*	GAATTC CTTAAG	G↓AATTC CTTAA↑G	GAATTC CTTAAG

TABLE 8-1 Recombinant DNA Is DNA with a New Sequence Formed by Joining Recognition Sequence Fragments from Different Sources

GAATTC will base pair with a sequence that runs in the opposite direction. EcoRI cleaves both DNA strands between the G and the A. When the two DNA fragments separate, they contain single-stranded complementary ends called *sticky ends*.

Each restriction enzyme name begins with the first three letters of the bacterium that produces it. This is illustrated in Table 8-1.

In 1972, David Jackson, Robert Symons, and Paul Berg generated recombinant DNA molecules. They allowed the sticky ends of the fragments to base pair with each other, and they covalently joined the fragments with the enzyme DNA ligase. The enzyme *DNA ligase* links the two sticky ends of the DNA molecules at the point of union. In 1973, Stanley Cohen and Herbert Boyer constructed the first recombinant plasmid capable of being replicated within a bacterial host. A *plasmid* is a circular DNA molecule that a bacterium can replicate without a chromosome.

In 1975, Edwin M. Southern developed procedures for detecting specific DNA fragments so that a particular gene could be isolated from a complex DNA mixture. This technique is called the *Southern blotting technique*. DNA fragments are separated by size with *agarose gel electrophoresis*. Gel electrophoresis takes advantage of the chemical and physical properties of DNA to separate the fragments. The phosphate groups in the backbone of DNA are negatively charged. This makes the DNA molecules attracted to anything that is positively charged. In gel electrophoresis, the DNA molecules are placed in an electric field so that they migrate toward the positive charge.

The DNA is placed in agarose, a polysaccharide, and placed in a tank of buffer. When electric current is applied, the DNA molecules migrate through the agarose gel, separate, and travel toward the positive poles of the electric fields. The DNA fragments migrate through the gel. The larger DNA fragments have a harder time moving than the smaller ones, so the small fragments travel farther through the gel. The agarose acts as a sieve, separating DNA based on size. Smaller fragments of DNA travel farther than larger fragments.

Still Struggling

Still having a problem with genetic engineering? *Genetic engineering* is the modification of an organism's genetic information. This is accomplished by recombinant DNA technology. *Recombinant DNA* technology takes a specific gene and inserts it into a vector to form a recombinant molecule. This molecule will make large amounts of the gene.

Artificial DNA: Putting Together the Pieces

Oligonucleotides, from the Greek word *oligo* meaning "few," are short pieces of DNA or RNA that are 2 to 30 nucleotides long. The ability to synthesize DNA oligonucleotides of a known sequence is incredibly important and useful. A DNA probe is used to analyze fragments of DNA. A DNA probe is a single-stranded fragment of DNA that recognizes and binds to a complementary section of DNA in a mixture of DNA molecules.

DNA probes can be synthesized, and DNA fragments can be prepared for use in molecular techniques such as the *polymerase chain reaction* (PCR). PCR is a technique that was developed by Kary Mullis in 1985. It produces large quantities of a DNA fragment without needing a living cell. Starting with one small piece of DNA, PCR can make billions of copies in a few hours using an instrument called a thermocycle. These large quantities of DNA can be analyzed easily.

PCR and DNA probes have been of great value to the areas of molecular biology, medicine, and biotechnology. Using these tools, scientists can detect the DNA associated with human immunodeficiency virus [HIV, the virus that causes acquired immune-deficiency syndrome (AIDS)], Lyme disease, chlamydia, tuberculosis, hepatitis, human papilloma virus (HPV) infection, cystic fibrosis, muscular distrophy, and Huntington's disease.

Gene Therapy: Makes You Feel Better

Gene therapy is a recombinant DNA process in which cells are taken from a patient, altered by inserting new DNA, and returned to the patient. A type of genetic surgery called *somatic gene therapy* is now possible. Cells of a person with a genetic disease are grown in a laboratory and altered with cloned DNA containing a normal copy of the defective gene. They could be reintroduced into the individual. If these cells become established, the expression of the normal genes may be able to cure the patient. This type of therapy is currently being used to treat diseases such as hemophilia and muscular dystrophy. The altered genes are not passed on to offspring.

In the early 1990s, gene therapy of this type was used to correct a deficiency of the enzyme *adenosine deaminase* (ADA). An immune-deficiency disease patient lacking the enzyme adenosine deaminase, an enzyme that destroys toxic metabolic by-products, had been treated. Some of the patient's lymphocytes were removed. Lymphocytes are a type of white blood cell that fight infection. The lymphocytes were given the *ADA* gene with the use of a modified retrovirus—which served as a *vector*—and placed back into the patient's body. Once established in the body, the cells with altered genes began to make the enzyme adenosine deaminase and alleviated the deficiency.

DNA Fingerprinting: Gotcha

DNA fingerprinting is an area of molecular biology that involves analyzing genetic material. It involves the use of restriction enzymes, which cut DNA molecules into pieces. When DNA samples obtained from different individuals are cut with the same restriction enzyme, the number and size of restriction fragments produced may be different. This difference provides the basis for DNA fingerprinting. The use of DNA fingerprinting depends on the presence of repeating base sequences. These sequences are called *restriction-fragment-length polymorphism* (RFLP) or the *RFLP pattern*, which is unique for every individual but is a combination of information from both biological parents.

RFLPs are a sort of molecular signature or fingerprint. In order to perform DNA fingerprinting, DNA must be taken from an individual. Samples can be taken from hair, white blood cells, skin, cheek cells, or other tissue. The DNA is taken from the cells and is broken down with enzymes. The fragments are separated with *electrophoresis*. The DNA fragments then are analyzed for RFLPs using DNA probes. An evaluation enables crime lab scientists (forensic pathologists) to compare a person's DNA with the DNA taken from a scene of a crime. This technique has a 99 percent degree of certainty that a suspect was at a crime scene. It also allows for tracking ancestry.

Industrial Application: Show Me the Money

Industrial applications of recombinant DNA technology include manufacturing protein products using bacteria, fungi, algae, or other cultured mammalian cells. The pharmaceutical industry is producing several medically important polypeptides using biotechnology such as insulin. Another example is *vaccines*. The hepatitis B vaccine is made up of a viral protein manufactured by yeast cells. Part of the virus is inserted into yeast which then replicates. The virus particle is then harvested and purified before being incorporated into the vaccine.

Agricultural Applications: Crops and Cows

Recombinant DNA and biotechnology have been used to increase plant growth by increasing the efficiency of the plant's ability to fix nitrogen. Scientists take genes for nitrogen fixation from bacteria and place the genes into plant cells. Because of this, plants can obtain nitrogen directly from the atmosphere. The plants can produce their own proteins without the need for bacteria. Another way to insert genes into plants is with a recombinant *tumor-inducing (TI) plasmid*. This is obtained from the bacterium *Agrobacterium tumefaciens*. These bacteria invade plant cells, and their plasmids insert chromosomes that carry the genes for tumor induction. An example of recombinant DNA used in livestock is the recombinant bovine growth hormone that has been used to increase milk production in cows by 10 percent.

U.S. farmers grow substantial amounts of genetically modified crops. About one-third of the corn crop and one-half of the soybean and cotton crops are genetically modified. Cotton and corn have become resistant to herbicides and insects. Soybeans have herbicide resistance and lower saturated fat content. Having herbicidal-resistant plants is important because many crop plants suffer stress when treated with herbicides. Resistant crops are not stressed by the chemicals that are used to control weeds.

Recombinant DNA Technology and Society: Too Much of a Good Thing

Genetically altering an organism raises scientific and philosophical questions. Recombinant DNA technology has had a positive impact on society, although there may be associated dangers with rDNA.

Concerns have been raised by the scientific community that genetically engineered microorganisms carrying dangerous genes may be released into the environment and cause widespread infection. Because of these worries, the federal government has established guidelines to regulate and limit the locations and types of experiments that are potentially dangerous.

Biomedical rDNA research has been regulated by the Recombinant DNA Advisory Committee (RAC) of the Natural Institutes of Health (NIH). The Food and Drug Administration (FDA) has principal responsibility in overseeing gene therapy research. The Environmental Protection Agency (EPA) and state governments have jurisdiction over field experiments in agriculture.

One of the biggest efforts in biotechnology has been the *Human Genome Project*, which began in 1990 and ended formally in 2001. The goal of this project was to determine the sequences of all human chromosomes. Advances such as this in biotechnology make genetic screening incredibly effective. Physicians will one day be capable of detecting genetic flaws in DNA long before the disease becomes manifested in a patient.

Another area of controversy is found in agriculture. Some scientists state that the release of recombinant organisms without risk assessment may disrupt the ecosystem. Viral nucleic acids, inserted into plants to make them resistant to viruses, may combine with the genome of an invading virus to make the virus even stronger. Genetically modified food may even trigger an allergic response in people or animals that consume them. As of this writing, obvious health or ecologic events have not been observed. However, owing to the consensus of the public, many food producers have stopped using genetically modified crops.

QUIZ

Fill in the blanks:

1. The modification of an organism's genetic information by changing its nucleic acid genome is called _____.

2. DNA fingerprinting is an area of molecular biology that involves analyzing the _____.

3. The use of DNA fingerprinting depends on the presence of repeating base sequences. These sequences are called _____.

4. Genetic engineering is accomplished by methods known as _____.

5. DNA with a new sequence formed by joining fragments from different sources is called _____.

Match the scientist:

    A. Stanley Cohen
    B. Dan Nathans
    C. Kary Mullis
    D. David Jackson
    E. Edwin M. Southern

6. This person was one of the scientists who had one of the fist breakthroughs leading to recombinant DNA by discovering microbial enzymes that make cuts into double-stranded DNA. _____

7. This person was one of the first scientists who generated recombinant DNA molecules. _____

8. This person was one of the scientists who constructed the first recombinant plasmid capable of being replicated within a bacterial host. _____

9. This scientist developed procedures for detecting specific DNA fragments to isolate a particular gene. _____

10. This scientist developed the polymerase chain reaction. _____

chapter 9

Classification of Microorganisms

Related microorganisms probably behavior similarly. A class is similar to a family because each microorganism in the class has similar characteristics. Therefore, knowing the class of a microorganism enables scientists to predict how that microorganism will behave. You'll learn the classification of microorganisms in this chapter and how to employ scientific techniques so that you can classify microorganisms.

CHAPTER OBJECTIVES

In this chapter, you will

- Be introduced to taxonomy
- Begin to learn taxonomy classification
- Understand taxonomy hierarchy
- Learn about the classification of organisms

A hotel lobby was filled with three family reunions all mingling together. We stood on a balcony overlooking the lobby, trying to determine who belonged to which family. At first, this seemed to be an insurmountable task; however, the more we viewed the people in the lobby, the more we noticed similarities among some of them. We could deduce family membership by looking at characteristics that seemed to belong to the same family and that probably belonged to a different family.

Scientists use a similar process to organize microorganisms into classes. A *class* is similar to a family because each microorganism in the class has similar characteristics. This is not to say that they are moms and dads or brothers and sisters. Scientists are saying that based on an unbiased observation, they have similar characteristics and probably are related to each other.

Taxonomy: Nothing to Do with the Internal Revenue Service

Organisms have traits that are similar to and different from those of other organisms. Scientists organize organisms into groups by developing a taxonomy. *Taxonomy* is the science of identifying and naming species based on genetic similarities, morphology, biochemistry, and physiology.

Scientists observe each organism, noting its characteristics. Organisms that have similar characteristics are presumed to have a natural relationship and therefore are placed in the same group. Classification tries to show this natural relationship.

Taxonomy has three components:

- *Classification.* The arrangement of organisms into groups based on similar characteristics, evolutionary similarity, or common ancestry. These groups are also called *taxa*.
- *Nomenclature.* The name given to each organism. Each name must be unique and should depict the dominant characteristic of the organism.
- *Identification.* The process of observing and classifying organisms into a standard group that is recognized throughout the biological community.

Taxonomy is a subset of systemics. *Systemics* is the study of organisms in order to place organisms having similar characteristics into the same group. Using techniques from other sciences such as biochemistry, ecology, epidemiology, molecular biology, morphology, and physiology, biologists are able to identify the characteristics of an organism.

Benefits of Taxonomy

Taxonomy organizes large amounts of information about organisms whose group members share many characteristics. Taxonomy allows scientists to make predictions and design a hypothesis for future research based on the knowledge of similar organisms. A *hypothesis* is a possible explanation for an observation that is tested through experimentation and testing.

If a relative of an organism has the same properties, that organism also may have the same characteristics. Taxonomy puts microorganisms into groups with precise names, enabling microbiologists to communicate with each other in an efficient manner. Taxonomy is indispensable for the accurate identification of microorganisms. For example, once a microbiologist or epidemiologist identifies a pathogen, physicians know the proper treatment that will cure the patient unless the pathogen becomes resistant.

Nomenclature of Taxonomy: Name Calling

In the mid-1700s, Swedish botanist Carl Linnaeus was credited as one of the first scientists to develop a taxonomy for living organisms. It is for this reason that he is known as the "father of taxonomy." Linnaeus' taxonomy grouped living things into two kingdoms: plants and animals.

By the 1900s, scientists had discovered microorganisms that had characteristics that were dramatically different from those of plants and animals. Therefore, Linnaeus' taxonomy needed to be enhanced to encompass microorganisms.

In 1969, Robert H. Whittaker, working at Cornell University, proposed a new taxonomy system that consisted of five kingdoms called *Whittaker's five-kingdom taxonomy* (Fig. 9-1). These kingdoms are monera, protista, plantae (plants), fungi, and animalia (animals). *Monera* are organisms that lack a nucleus and membrane-bounded organelles, such as bacteria. *Protista* are organisms that have either a single cell or no distinct tissues and organs, such as protozoa. This group includes unicellular eukaryotes and algae. *Fungi* are organisms that use absorption to acquire food. These include multicellular fungi and single-cell yeast. Animalia and plantae include only multicellular organisms.

Scientists widely accepted Whittaker's taxonomy until 1977, when Carl Woese, in collaboration with Ralph S. Wolfe at the University of Illinois, proposed a new six-kingdom taxonomy. This came about with the discovery of *archaea*, which are prokaryotes that live in oxygen-deprived environments.

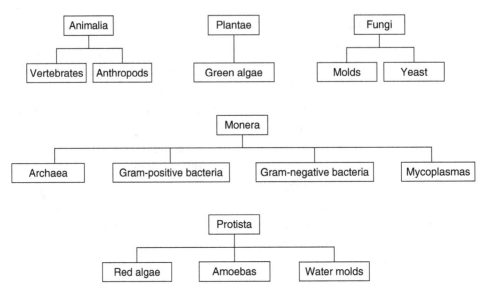

FIGURE 9-1 • Whittaker's five-kingdom taxonomy.

Before Woese's six-kingdom taxonomy, scientists grouped organisms into eukaryotes—animals, plants, fungi, and one-cell microorganisms (paramecia)—and prokaryotes (microscopic organisms that are not eukaryotes).

Woese's five-kingdom taxonomy in 1969 consisted of

- Eubacteria (has rigid cell wall)
- Protista (unicellular eukaryotes and algae)
- Fungi (multicellular forms and single-cell yeasts)
- Plantae
- Animalia

By studying the ribosomal RNA (rRNA) sequences in prokaryotic cells, Woese determined that archaebacteria and eubacteria are two groups.

Woese used three major criteria to define his six kingdoms. These are

- *Cell type.* Eukaryotic cells (cells having a distinct nucleus) and prokaryotic cells (cells not having a distinct nucleus).
- *Level of organization.* Organisms that live in a colony or alone and one-cell organisms and multicell organisms.
- *Nutrition.* Ingestion (animal), absorption (fungi), or photosynthesis (plants).

In the 1990s, Woese studied rRNA sequences in prokaryotic cells (archaebacteria and eubacteria), proving that these organisms should be divided into two distinct groups. Today, organisms are grouped into three categories called *domains* that are represented as bacteria, archaea, and eukaryotes.

The domains are placed above the phylum and kingdom levels. The term *archaebacteria* (from the Greek word *archaio*, meaning "ancient") refers to the ancient origin of this group of bacteria that appear to have diverged from eubacteria.

The evolutionary relationship among the three domains is

- Domain bacteria (eubacteria)
- Domain eukarya (eukaryotes)

Different classifications of organisms include

- Bacteria
  ○ Eubacteria
- Archaebacteria
  ○ Archaea
- Eukarya
  ○ Protista
  ○ Fungi
  ○ Plantae
  ○ Animalia

The three domains are archaea, bacteria, and eukarya (Fig. 9-2):

- *Archaea* lack peptidoglyan acid in the cell walls.
- *Bacteria* have a cell wall composed of peptidoglycan and muramic acid. Bacteria also have membrane lipids with ester-linked, straight-chained fatty acids that resemble eukaryotic membrane lipids. Most prokaryotes are bacteria. Bacteria also have plasmids, which are small, double-stranded DNA molecules that are extrachromosomal.
- *Eukarytes* are of the domain eukarya and have a defined nucleus and membrane-bound organelles and have linear chromosomes.

Taxonomic Rank and File

A taxonomy has an overlapping hierarchy that forms levels of *rank* or *category* similar to an organization chart. Each rank contains microorganisms that have similar characteristics. A rank also can have other ranks that contain microorganisms.

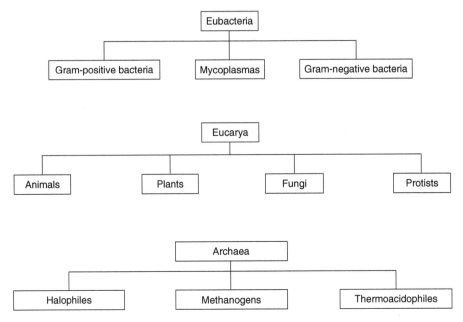

FIGURE 9-2 • Three-domain taxonomy.

Microorganisms that belong to a lower rank have characteristics that are associated with a higher rank to which the lower rank belongs. However, characteristics of microorganisms of a lower rank are not found in microorganisms that belong to the same higher rank as the lower-rank microorganism.

Microbiologists use a *microbial taxonomy* (Fig. 9-3), which is different from what biologists, who work with larger organisms, use. Microbial taxonomy is commonly

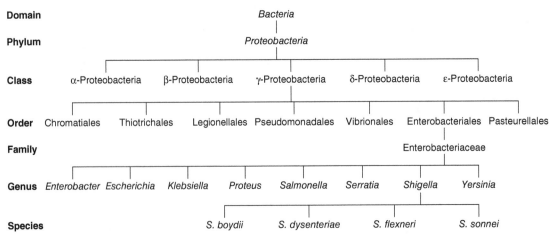

FIGURE 9-3 • Microbiologists use a microbial taxonomy.

called *prokaryotic taxonomy*. The widely accepted prokaryotic taxonomy is *Bergey's Manual of Systematic Bacteriology*, first published in 1923 by the American Society for Microbiology. David Bergey was chairperson of the editorial board.

In the taxonomy of prokaryotes, the most commonly used rank (in order from most general to most specific) is

Domain

    Kingdom

        Phyla

           Class

               Order

                  Family

                     Genus

                        Species

The basic taxonomic group in microbial taxonomy is the *species*. Taxonomists working with higher organisms define their species differently than microbiologists. Prokaryotic species are characterized by differences in their phenotype and genotype. *Phenotype* is the collection of visible characteristics and the behavior of a microorganism. *Genotype* is the genetic makeup of a microorganism.

Prokaryotic species are collections of strains that share many properties and differ dramatically from other groups or strains. A *strain* is a group of microorganisms that share characteristics that are different from microorganisms in other strains. Each microorganism within a strain is considered to have descended from the same microorganism.

For example, *Biovars* are a species that contains strains characterized by differences in their biochemistry and physiology. *Morphovars* are also a species whose strains differ morphologically and structurally. *Serovars* are another species that has strains that are characterized by distinct antigenic properties (substances that stimulate the production of antibodies).

Microbiologists use the genus of the taxonomy to name microorganisms, which you learned in Chapter 1. Microorganisms are given a two-part name. The first part is the Latin name for the genus. The second part is the epithet. Together these parts uniquely identify the microorganism. The first part of the name is always capitalized, and the second part of the name is always lowercase. Both parts are italicized.

For example, *Escherichia coli* is a bacterium that is a member of the *Escherichia* genus and has the epithet *coli*. Usually, after first mention in a piece of writing, the name is abbreviated as *E. coli*. However, the abbreviation maintains the same style as the full name (uppercase, lowercase, italic).

Classification: All Natural

A taxonomy is based on scientists' ability to characterize organisms into a classification system. The most widely used classification system is called the *natural classification*. The natural classification requires that an organism be grouped with organisms that have the same characteristics.

In the mid-eighteenth century, Linnaeus developed the first natural classification using anatomic characteristics of organisms. Other natural classifications use classical characteristics to group organisms. These characteristics are

- *Morphologic.* Morphologic characteristics classify organisms by their structure, which normally remains the same in a changing environment and is a good indication of phylogenetic relatedness.
- *Ecologic.* Ecologic characteristics classify organisms by the environment in which they live. For example, some microorganisms live in various parts of the human intestine, and others live in marine environments. Ecologic characteristics include the temperature, pH, and oxygen requirements of an organism, as well as an organism's life cycle.
- *Genetic.* Genetic characteristics classify organisms by the way in which they reproduce and exchange chromosomes. For example, eukaryotic organisms reproduce sexually by conjugation, where two cells come together and exchange genetic material. Prokaryotic organisms do not reproduce sexually and instead use binary fission to reproduce. This typically occurs between strains of prokaryotes, if their genomes are dissimilar, but rarely between genera.

In the early 1990s, T. Cavalier-Smith developed the two-empire and eight-kingdom taxonomy based on phentic and phylogenetic characteristics. *Phentic* characteristics are the physical characteristics of an organism, and they are determined using a process called *numerical taxonomy*. Numerical taxonomy is a phentic classification based on physical measurements of an organism. *Phylogenetic* characteristics are the evolutionary relationships among organisms.

The two empires are bacteria and eukaryota. The bacteria domain contains two kingdoms. These are eubacteria and archaeobacteria. The eukaryota empire contains six kingdoms, as shown in Table 9-1.

TABLE 9-1 Cavalier-Smith Two-Empire and Eight-Kingdom Taxonomy	
Empire	**Kingdom**
Bacteria	Eubacteria: Large group of bacteria that have rigid cell walls.
	Archaeobacteria: Nonrigid cell walls.
Eukaryota	Archezoa: Primitive one-cell eukaryotes.
	Chromista: Photosynthetic organisms that have chloroplasts.
	Plantae: Photosynthetic organisms that have chloroplasts in the cytoplasmic matrix.
	Fungi: Absorb nutrients.
	Animalia: Ingest nutrients.
	Protozoa: Single-cell organism.

QUIZ

Fill in the blanks:

1. This scientist is referred to as the father of taxonomy. _____

2. In 1969, Robert Whittaker proposed a new taxonomy that consisted of _____.

3. According to Whittaker's five-kingdom taxonomy, bacteria would be in the kingdom _____.

4. According to Whittaker's five-kingdom taxonomy, green algae would in the kingdom _____.

5. According to the three-domain taxonomy, fungi would be in the domain _____.

Match the scientist:

 A. T. Cavalier-Smith
 B. Robert H. Whittaker
 C. Carl Woese
 D. Carl Linnaeus
 E. Ralph S. Wolfe
 F. David Bergey

6. This scientist was one of the first scientists to develop a taxonomy for living organisms. _____

7. This scientist used a five-kingdom taxonomy for classifying organisms. _____

8. These two scientists used six kingdoms for classifying organisms. _____ and _____

9. This scientist chaired the editorial board for the *Manual of Systematic Bacteriology*. _____

10. This scientist developed the two-empire and eight-kingdom taxonomy. _____

The Prokaryotes: Domains Archaea and Bacteria

In this chapter, you'll learn about the major types of bacteria. Bacteria are prokaryotes and are divided into four divisions called *phyla* based on characteristics of their cell walls. The phyla are further subdivided into *sections* based on other characteristics, such as Gram stain reaction, motility, oxygen requirements, and shape. The sections are divided into *genera*. By the end of this chapter, you will understand these terms as the basis for a family tree for bacteria.

CHAPTER OBJECTIVES

In this chapter, you will

- Learn about the domain archaea
- Examine microaerophilic gram-negative bacteria
- Be introduced to the genus *Spirochete*
- Learn about the genera *Rickettsia* and *Chlamydia*

You have bacteria in you. Hearing these words is probably alarming because you imagine the time that you had a sore throat or other bacterial infection that made you feel miserable, sending you to the doctor to get a prescription for an antibiotic.

Yet not all bacteria cause disease. Some bacteria cause disease, whereas other bacteria are necessary for you to digest certain foods. Antibiotics don't know the difference between good and bad bacteria, and, as a result, antibiotics kill the good with the bad, which can be problematic because you need the good bacteria. The doctor probably told you to eat yogurt or take a probiotic capsule. Both contain good bacteria and replace the good bacteria killed by the antibiotics.

Archaea

Archaea can exist in very hot and very cold environments, making them resilient microorganisms that can adapt to environments that destroy other bacteria. For example, archaea can survive and grow in an oxygen-free environment (anaerobic) and in a high-salt (hypersaline) environment.

There are three ways that microbiologists identify archaea. Archaea

- Have a unique sequence of ribosomal RNA (rRNA)
- Have cell walls that lack peptidoglycan. The cell walls of most bacteria contain peptidoglycan
- Have a unique membrane lipid that has a branched chain of hydrocarbons connected to glycerol ester links. The membrane lipid of most bacteria has glycerol connected to fatty acids by ester bonds

Unfortunately, two of the more common techniques used to identify bacteria are not very useful in identifying archaea. You'll recall from Chapter 4 that microbiologists identify bacteria by using Gram's stain. A bacterium is either gram-positive or gram-negative. However, archaea cell walls do not contain peptidoglycan and cannot be stained, which makes Gram's stain useless when trying to identify archaea.

The shape of a bacterium is another common way microbiologists identify bacteria. Many bacteria have a distinctive appearance. However, archaea are *pleomorphic*, which means that they can have various shapes. Sometimes archaea are spiracle-shaped, spiral, lobed, plate-shaped, or irregularly shaped.

Archaea also have various types of metabolism. Some archaea are organotrophs, whereas others are autotrophs. Archaea also break down (catabolize) glucose for energy in various ways. It is these variations that enable archaea to survive in environments that are fatal to other bacteria.

The Archaea Clan

Archaea are not bacteria, but they can be organized into subgroups. Microbiologists use one of two subgroup classifications for archaea. One classification method divides archaea into five subgroups. These are

- *Methanogenic archaea.* A single-celled archaea is anaerobic and even low levels of O_2 can be toxic. Some reduce CO_2 into methane as waste in the presence of H_2.
- *Sulfate reducers.* Archaea reduce sulfate to hydrogen sulfide to obtain energy in the presence of air.
- *Extreme halophiles.* Archaea that live in an extremely salty environment.
- *Cell wall-less archaea.* Archaea that do not have a cell wall.
- *Extremely thermophilicsulfur metabolizers.* Archaea that need sulfur at high temperatures for growth.

The other method used to organize archaea into subgroups is that used in *Bergey's Manual of Systematic Bacteriology* that you learned about in Chapter 9; it consists of two branches (phyla). These are

- *Phylum Crenarchaeota.* Archaea that are within the phylum *Crenarchaeota* branch are anaerobes that grow in a sulfur-enriched soil or water environment that is at a temperature between 88 and 100°C and has a pH between 0 and 5.5. Extremely thermophilic sulfur metabolizers are within the phylum *Crenarchaeota* subgroup.
- *Phylum euryarchaeota.* The phylum euryarchaeota branch consists of the following five major groups:
  - *Methanogenic archaea.* Methanogenic archaea, the largest group of the phylum *Curyarchaeota*, are anaerobic archaea that synthesize organic compounds in a process called *methanogenesis*, which produces methane (Fig. 10-1). They also use inorganic sources (autotrophic) such as H_2 and CO_2 for growth. Methanogenic archaea thrive in swamps, hot springs, and fresh water, as well as in marshes. They digest sludge and transform undigested food, in animal intestines and in the rumen of a ruminant, into methane. A *ruminant* is a herbivore that has a stomach that is divided into four compartments. The *rumen* is the expanded

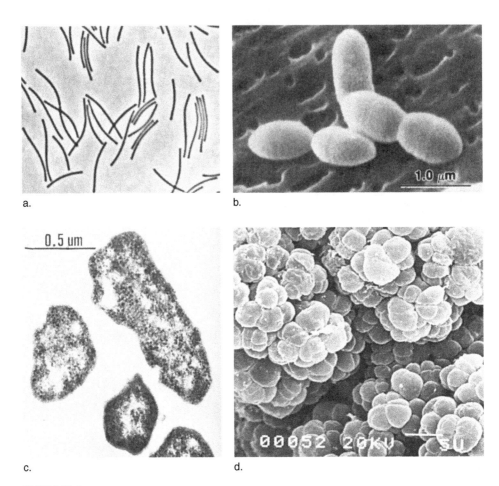

FIGURE 10-1 • Selected Methanogens. (a) *Methanospirillumhungatei*; phase contrast (×2,000). (b) *Methanobrevibactersmithii*. (c) *Methanogeniummarisnigri*; electron micrograph (×45,000). (d) *Methanosarcinamazei*; SEM. Bar = 5 μm. (From Prescott et al., *Microbiology*, 6th ed., McGraw-Hill, 1996.)

upper compartment of the stomach that contains regurgitated and partially digested food called a*cud*. Methanogenic archaea transform regurgitated and partially digested food into methane (CH_4), which is a clean-burning fuel. For example, a cow can belch up to 400 liters of methane a day. Sewage treatment plants also use methanogenic archaea to transform organic waste into methane. Although methane is a source of energy, it is also a cause for the greenhouse effect. Methanogens are further organized into four orders. These are methanobacteriales, methanococcales, methanomicrobiales, and methanosarcinales. Examples of methanogenic archaea include *Methanospirillumhungatei*, *Methanobrevibactersmithii*, *Methanogeniummarisnigri*, and *Methanosarcinamazei*.

○ *Extreme halophiles.* Extreme halophiles, also known as *halobacteria*, absorb nutrients from dead organic matter in the presence of oxygen (*aerobic chemoheterotrophs*) (Fig. 10-2). They require proteins, amino acids, and other nutrients for growth in a high concentration of sodium chloride. Extreme halophiles can be motile or nonmotile and are found in salt lakes and in salted fish. Extreme halophiles turn lakes and fish red when they exist in abundance.

○ *Halobacteriumsalinarium.* *H. salinarium* is an archaea that acquires energy through photosynthesis. However, it is able to do so without the need for chlorophyll or bacteriochlorophyll. *H. salinarium* synthesize the protein *bacteriorhodopsin*, which is deep purple in color under high-intensity lighting in a low-oxygen environment.

○ *Thermophilicarchaeons.* Thermophilicarchaeons are known as *thermoplasma* and grow hot (55 to 59°C), acidic (pH of 1 to 2) refuse piles of coalmines that contain iron pyrite (Fig. 10-3). These refuse piles become hot and acidic as chemolithotropic bacteria oxidize iron pyrite into sulfuric acid. Thermophilicarchaeons lack a cell wall.

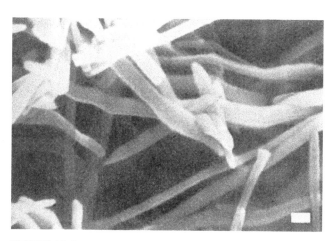

FIGURE 10-2 • *Halobacteriumsalinarium.* A young culture that has formed long rods; SEM. Bar = 1 μm. (From Prescott et al., *Microbiology*, 6th ed., McGraw-Hill, 1996.)

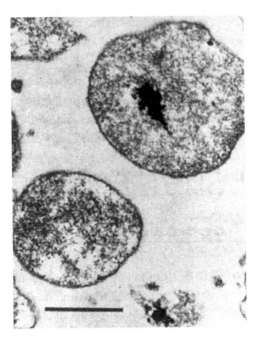

FIGURE 10-3 • *Thermoplasma.* Transmission electron micrograph. Bar = 0.5 μm. (From Prescott et al., *Microbiology,* 6th ed., McGraw-Hill, 1996.)

- *Sulfate-reducing archaea.* Sulfate-reducing archaea are known as *archaeoglobi;* they extract electrons from various donors to reduce sulfur to sulfide in an environment that is approximately 83°C, such as near marine hydrothermal vents (underwater hot springs). Sulfate-reducing archaea are gram-negative and are shaped as irregular spheres (coccoid cells).

Still Struggling

Having problems understanding archaea? Archaea can live in very hot or cold environments. *Methanogenic archaea* are single-celled archaea that produce methane and carbon dioxide. *Sulfate reducers* are archaea that reduce sulfate to hydrogen sulfide to obtain energy. *Extreme halophiles* are archaea that live in an extremely salty environment. *Cell wall-less archaea* are archaea that do not have a cell wall and have no specific shape. Exremely *thermophilicsulfur metabolizers* are archaea that need sulfur to live and survive.

Aerobic/Microaerophilic, Motile, Helical/Vibroid, Gram-Negative Bacteria

Another kind of prokaryote is the aerobic/microaerophilic, motile, helical/vibroid, gram-negative bacterium. This is a mouthful to say, but the name describes characteristics of this group of prokaryotic bacteria.

Aerobic/microaerophilic means that bacteria within this group require small amounts of oxygen to grow. *Motile* implies that the bacterium is self-propelled, using flagella at one or both poles to move in a corkscrew motion. *Helical/vibroid* indicates that the bacterium takes the shape of a spiral (helical) or a curved rod (vibroid). *Gram-negative* means that when the bacterium is identified using Gram's stain, it loses the violet stain when rinsed and appears red or pink. Bacterium has a thin layer of peptidoglycan in the cell wall to which the stain attaches.

Aerobic/microaerophilic, motile, helical/vibroid, gram-negative bacteria thrive in soil and are found on the roots of plants such as the *Azospirillum*, where they improve the plant's nutrient uptake. Bacteria within this group are also found in both fresh and stagnant water.

Some aerobic/microaerophilic, motile, helical/vibroid, gram-negative bacteria such as *Campylobacter fetus* and *C. jejuni* cause diseases (pathogenic). *C. fetus* causes spontaneous abortion in domestic animals. *C. jejuni* causes inflammation of the digestive tract (enteritis) as a result of food-borne intestinal diseases. Another common aerobic/microaerophilic, motile, helical/vibroid, gram-negative bacterium that is pathogenic is *Helicobacter pylori*, which causes gastric ulcers in humans (Fig. 10-4).

Still Struggling

Having problems understanding types of bacteria? *Microaerophilic bacteria* require a very small amount of oxygen to live and grow. *Motile bacteria* are self-propelled. *Helical bacteria* are spiral-shaped. *Vibroid bacteria* are curved rods. *Cocci bacteria* are round. *Bacilli bacteria* are rod-shaped. *Staph bacteria* are organized in clusters. *Strep bacteria* are organized in chains.

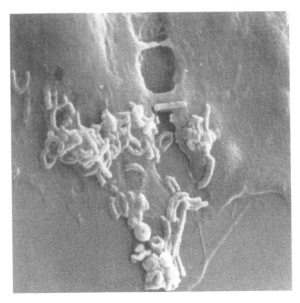

FIGURE 10-4 • Peptic ulcer disease. Scanning electron micrograph (×3,441) of *Helicobacter pylori* adhering to gastric cells. (From Prescott et al., *Microbiology,* 6th ed., McGraw-Hill, 1996.)

Gram-Negative Aerobic Rods and Cocci

Bacteria that are members of the gram-negative aerobic rods and cocci group include many bacteria that cause disease in humans and bacteria that are important to industry and the environment. There are 11 bacteria in this group:

- *Pseudomonads.* Pseudomonads are rod-shaped bacteria with polar flagella, which give the bacteria mobility. They need oxygen to grow and obtain energy by decomposing organic material. Pseudomonads are found in soil and freshwater and marine environments. *Pseudomonas aeruginosa* is a pathogenic bacterium that infects the urinary tract and wounds in humans. It also causes infections in burn injuries.

- *Legionella. Legionella pneumophila* is a bacterium identified in 1976 when it infected and killed members of the American Legion at their convention in Philadelphia. Infection caused by the *L. pneumophila* is commonly referred to as *Legionnaire's disease. L. micdadei* is the bacterium that infects lungs and causes a strain of pneumonia commonly called *Pittsburgh pneumonia.*

- *Moraxella lacunata. M. lacunata* is an egg-shaped (coccobacilli) bacterium that can infect the membrane that lines *eyelids, called the conjuntiva,* causing a condition known as *conjunctivitis* ("pink eye").

- *Neisseria. Neisseria* are double-spherical (diplococcal) bacteria that can live with or without oxygen (anaerobic) and usually are found on mucous membranes in humans. One type of *Neisseria* called *N. gonorrhoeae* causes the sexually transmitted disease gonorrhea (Fig. 10-5). Another type is *N. meningitidis (N. meningitis)*, the bacterium that infects the mucous membranes of the nose and throat (*nasopharyngeal*), causing a sore throat. *N. meningitidis* can cause meningitis if the bacterium enters blood and cerebral spinal fluid, where it can infect the protective covering (meninges) of the brain and spinal cord. Meningitis is inflammation of the meninges.

- *Brucella. Brucella* are very small coccobacilli that cannot move themselves (nonmotile) and cause brucellosis or undulant fever—daily episodes of fever and chills. *Brucella* multiply in white blood cells.

- *Bordetella pertussis. B. pertussis* are the bacteria that cause pertussis, which is better known as *whooping cough. B. pertussis* are rod-shaped and nonmotile.

- *Franeisella. Franeisella tularensis* is a gram-negative coccobacillus bacterium that lives in contaminated water or wild game. When such water or wild game is ingested, the bacteria infects the lymph nodes (lymphadenopathy), causing a disease called *tularemia*, which is commonly known as *rabbit fever* or *deer-fly fever. F. tularensis* also can be inhaled during the skinning of an infected animal or enter through a lesion in the skin.

- *Agrobacterium. Agrobacterium tumefaciens* is a bacterium that causes tumor-like growths on plants called *crowngull*.

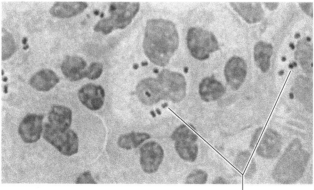

Diplococci

FIGURE 10-5 · *Neisseria gonorrhoeae.* The diploccci are often within white blood cells (×1,000). (From Prescott et al., *Microbiology,* 6th ed., McGraw-Hill, 1996.)

- *Acetobacter* and *Gluconobacter*. *Acetobacter* and *Gluconobacter* are bacteria that synthesize ethanol to vinegar (acetic acid) and are used in the food industry to make vinegar.

- *Spirochaeta*. *Spirochaeta*, which means "coiled hair," are gram-negative chemoheterotrophs (organisms that break down organic compounds for energy). These organisms are skinny, long, flexible bacteria that move by means of axial filaments (*endoflagella*) that encircle the cell. The spinning of these filaments causes movement that resembles a corkscrew. Examples of spirochetes include *Treponemapallidum* (Fig. 10-6) and *Borreliaburgdorferi* (Fig. 10-7).

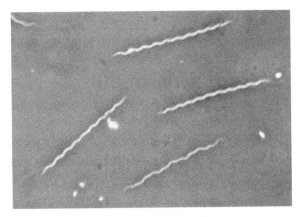

FIGURE 10-6 · *Treponemapallidum* (×1,000). (From Prescott et al., *Microbiology*, 6th ed., McGraw-Hill, 1996.)

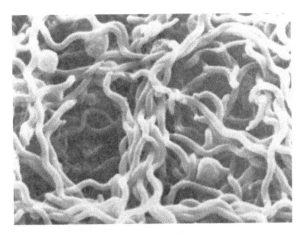

FIGURE 10-7 · Lyme Disease. The etiological agent is the spirochete *Borreliaburgdorferi*. (From Prescott et al., *Microbiology*, 6th ed., McGraw-Hill, 1996.)

Facultative Anaerobic Gram-Negative Rods

Facultative anaerobic gram-negative rods are a group of bacteria that have a rod shape and can live without the presence of oxygen, or, when oxygen is present, can carry out metabolism aerobically. These bacteria are gram-negative. There are three prominent members of the facultative anaerobic gram-negative rods group.

- *Enterics.* Enterics (Enterobacteriaceae) are small bacteria that are found in the intestinal tracts of animals and humans (*intestinal flora*) and most have flagella all over their surfaces (*peritirichous flagella*) to move about. Enterics ferment glucose and produce carbon dioxide and other gases. The word *enteric* means "pertaining to the intestines." More predominant enterics include
  - *Escherichia coli.* E. *coli* is an example of an enteric bacteria that makes up some of the normal flora in the human intestines but can cause infection if it enters other parts of the body (e.g., through the ingestion of water that is contaminated with fecal matter) (Fig. 10-8).
  - *Shigella.* Shigella causes bacillary dysentery or shigellosis, more commonly known as *traveler's diarrhea.*
  - *Salmonella.* Salmonella is a group (genera) of enteric bacteria that has members that can infect humans. Species include *Salmonella typhi,* which causes typhoid fever, and S. *choleraesuis* and S. *enteritides,* both of which are food-borne pathogens that cause salmonellosis, a type of food poisoning.
  - *Klebsiella.* The genus *Klebsiella* contains bacteria that cause pneumonia.
  - *Erwinia.* Erwinia is a bacterium that infects plants and causes what is commonly called *soft root rot.*
  - *Enterobacter.* The genus *Enterobacter* consists of the species E. *cloace* and E. *aerogenes.* Both these organisms cause urinary tract infections and nosocomial (hospital) infections in individuals with a weakened immune system.
  - *Serratiamarcescens.* S. *marcescens* are found on catheters and instruments that are allegedly sterile; this bacterium causes urinary and respiratory infections.
  - *Yersinia pestis.* Y. *pestis* caused the bubonic or black plague that ravaged Europe during the Middle Ages. It begins by causing abscesses of lymph nodes and then produces pneumonia-like symptoms when it reaches the lungs, which is called *pneumonic plague.*

- *Vibrios.* Vibrios are facultative anaerobic gram-negative comma-shaped bacteria that inhibit aquatic environments, and some also live in the intestinal tracts of animals and humans. There are two important species of vibrios. These are
  - *Vibriocholerae. V. cholerae* are the bacteria that cause cholera (Fig. 10-9). Signs and symptoms include abdominal pain and watery diarrhea (diarrhea appears like rice water).
  - *Vibrio parahaemolyticus. V. parahaemolyticus* are the bacteria that cause inflammation and irritation of the stomach and intestine—better known as *gastroenteritis*—when contaminated shellfish is ingested raw or undercooked.

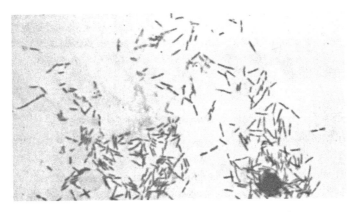

FIGURE 10-8 • *E. coli*, Gram stain (×500). (From Prescott et al., *Microbiology*, 6th ed., McGraw-Hill, 1996.)

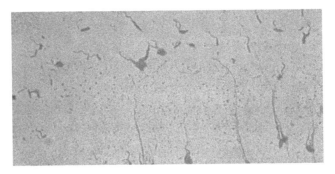

FIGURE 10-9 • *V. cholerae*. Curved rods with polar flagella (×1,000). (From Prescott et al., *Microbiology*, 6th ed., McGraw-Hill, 1996.)

- *Pasteurella-Haemophilus. Pasteurella-Haemophilus* are very small, facultative, anaerobic, gram-negative, rod-shaped bacteria that are named for Louis Pasteur. Examples include
  - *Pasteurella.* Theses bacteria cause blood poisoning (septicemia) in cattle and chickens (fowl cholera) and pneumonia in various other animals.
  - *P. multocida*, the species that was identified by Louis Pasteur, is transmitted to humans by a dog or cat bite.
  - *Haemophilus. Haemophilus* (which means "blood loving") are bacteria that live in mucous membranes of the upper respiratory tract, mouth, intestinal tract, and vagina.
  - *H. ducreyi* is the bacterium that causes *chancroid*, which is an infectious venereal ulcer.
  - *H. aegyptiusis* is the species that causes acute conjunctivitis, or pink eye.

Anaerobic Gram-Negative Cocci and Rods

Anaerobic gram-negative cocci and rods can live in anaerobic conditions and are nonmotile. They do not form *endospores*, which are small spores that develop inside the bacteria as a resistant survival form of the bacteria. *Anaerobic gram-negative cocci* are spherical in shape and form a single chain or are clustered. *Veillonella* are common anaerobic gram-negative cocci that are found between the teeth and on gums. *Veillonella* is the cause of abscesses of teeth and gums.

 Anaerobic gram-negative rods are called *bacteroides* and are members of the Bacteriodaceae family of bacteria that live in the intestinal tract of humans. Bacteroides can cause *peritonitis*, which is inflammation of the peritoneum owing to infection. Another kind of anaerobic gram-negative rod is *Fusobacterium*. These are long, slender rods that live in the gingival crevices of teeth and cause *gingivitis*, which is a gum infection.

Rickettsias and Chlamydias

Rickettsias and chlamydias are intracellular parasites that need a host in order to reproduce and therefore enter the cell of a host. These bacteria were once thought to be viruses that invaded cells. They are classified as bacteria because they have bacterial cell walls and contain DNA and RNA, which is not the case with a virus. *Rickettsias* and chlamydias have no means of mobility because they lack flagella. They are also gram-negative.

Rickettsias are small rod-shaped or spherical bacteria of the genus *Rickettsia* that live in the cells of ticks, lice, fleas, and mites (arthropods) and can be transmitted to humans when they are bitten by arthropods, causing rickettsial disease. *Rickettsial disease* causes capillaries to become permeable, resulting in a rash. Rickettsias reproduce by *binary fission*, where a cell wall forms across the cell, and the two halves separate to become individual cells.

Common rickettsias include

- *Rickettsia prowazekii. R. prowazekii* is transmitted by lice and causes *endemic typhus*.
- *R. rickettsii. R. rickettsii* is transmitted by ticks and causes *Rocky Mountain spotted fever*.
- *R. tsutsugamushi. R. tsutsugamushi* is transmitted by arthropods and causes *scrub typhus* that presents with fever, rash, and inflammation of the lymph nodes.
- *Coxiellaburnetti. C. burnetti* is transmitted by aerosols or contaminated milk and causes *Q fever*, which is similar to pneumonia.
- *Bartonellabacilliformis. B. bacilliformis* is transmitted by arthropods and causes a wartlike rash called *oroyafever*.

Chlamydias are very small spherical or coccoid bacteria of the genus *Chlamydia* that are nonmotile and can be transmitted via person-to-person contact or by airborne respiration. Chlamydias are not transmitted by arthropods. They are so small that they multiply in host cells. There are three species of *Chlamydia*:

- *C. trachomatis* causes *trachoma*, which is a common cause of blindness, and the nongonorrhea sexually transmitted disease *urethritis* (inflammation of the urethra).
- *C. pneumonia* causes a mild form of pneumonia in adolescence.
- *C. psittaci* causes *psittacosis*.

Mycoplasmas

Mycoplasmas are very small facultative anaerobic bacteria (some are obligate anaerobes) of the genus *Mycoplasma* that can take on many shapes (*pleomorphic*) and once were thought to be viruses because they lack a cell wall. However, they have a cell membrane, DNA, and RNA, which distinguishes them from viruses.

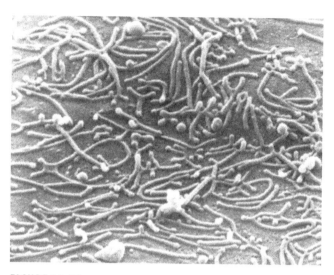

FIGURE 10-10 · *M. pneumonia,* SEM (×62,000). (From Prescott et al., *Microbiology,* 6th ed., McGraw-Hill, 1996.)

Mycoplasmas also can resemble fungi because some mycoplasmas produce filaments that are seen commonly in fungi. It is these filaments that led scientists to name the genus *Mycoplasma* (*myco* means "fungus").

Many mycoplasmas are unable to move by themselves because they do not have flagella, but some are able to glide on a wet surface.

Two of the more common types of mycoplasmas are

- *M. pneumoniae* is the cause of atypical pneumonia, commonly referred to as *walking pneumonia* (Fig. 10-10).

- *Ureaplasma urealyticum* is a bacterium that is found in urine and causes urinary tract infection.

Gram-Positive Cocci

Within the gram-positive cocci section are two important genera. These are *Staphylococcus* and *Streptococcus*, and each has an important role in medicine.

- *Staphylococcus. Staphylococcus* bacteria have a grapelike cluster arrangement and grow in environments of high solute concentration and low moisture. *Osmotic pressure* is the pressure required to prevent the net flow of water by osmosis. Infections caused by *Staphylococcus* bacteria typically are called *staph infections*. Here are the commonly found *Staphylococcus* bacteria:

- ○ *Staphylococcus aureus. S. aureus* is a bacterium that forms yellow-pigmented colonies that grow with (aerobically) or without oxygen (anaerobically) (Fig. 10-11). *S. aureus* is the cause of *toxic shock syndrome*, which results in high fever, vomiting, and sometimes death. It produces *enterotoxins*, which affect intestinal mucosa. *S. aureus* is also the cause of *boils* (skin abscess), *impetigo* (pus-filled blisters on the skin), *styes* (an infection at the base of an eyelash), *pneumonia, osteomyelitis, acute bilateral endocarditis* (inflammation of the internal membrane of the heart), and *scalded-skin syndrome* in very young children that causes skin to strip off (denude) owing to an exfoliative toxin. *S. aureus* is one of the major types of infections that occur in hospitals because it is resistant to many antibiotics such as penicillin. An infection with *S. aureus* is usually identified by the presence of an abscess.

- ○ *Staphylococcus epidermis. S. epidermis* is the frequent cause of *urinary tract infections* in the elderly and also causes *subacute bacterial endocarditis*, which is a chronic infection of the endocardium (the thin layer of connective tissue that lines the chambers of the heart) and heart valves. It enters the body through cracks in the skin or after surgical procedures.

- ○ *Staphylococcussaprophyticus. S. saprophyticus* causes *urinary tract infections*, usually in adolescent girls.

- • *Streptococcus. Streptococcus* bacteria appear as a paired or chained spherical gram-positive bacteria. Streptococci do not use oxygen, although most are

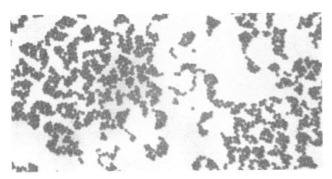

FIGURE 10-11 • *S. aureus*. Note the gram-positive spheres in irregular clusters. Gram stain (×1,000). (From Prescott et al., *Microbiology*, 6th ed., McGraw-Hill, 1996.)

aerotolerant. Few may be obligatelyanarobic. Infections caused by the *Streptococcus* bacteria generally are referred to as *strep infections*. Microbiologists classify *Streptococcus* bacteria in three ways:

- *The type of hemolysis (destruction) of red blood cells caused by the Streptococcus bacteria.* There are three types, characterized by
  - *α-Hemolytic group.* Incomplete *lysis* (destruction of the cell) within green pigment surrounding the colony.
  - *β-Hemolytic group.* Total lysis and a clear area around the colony.
  - *γ-Hemolytic group.* Absence of lysis. This group is of no clinical importance.
- *The Lancefield classification.* There are four groups:
  - *Group A streptococci.* Characterized by *S. pyogenes*, they secrete erythrogenic exotoxins responsible for *scarlet fever*.
  - *Group B streptococci.* Characterized by *S. agalactiae*, which is part of normal oral and vaginal flora, they cause *urogenital* (urinary and reproductive system) infections in females.
  - *Group C streptococci.* They cause animal diseases.
  - *Group D streptococci.* Characterized by *S. faecalis*, which is a normal part of oral and intestinal flora. Diseases of *S. faecalis* are endocarditis, urinary tract infections, and septicemia (blood poisoning).
- *Ungrouped streptococci.* There are two important kinds:
  - *Viridans streptococci.* Characterized by *S. viridans* and *S. salvarius*, which causes *subacute bacterial endocarditis*, and *S. mutans*, which causes a biofilm called *plaque* resulting in tooth decay.
  - *Pneumococcal streptococci.* Characterized by *S. pneumonia*, which causes *lobar pneumonia* and *otitis media* (middle ear infection).

Endospore-Forming Bacteria

The vast majority of endospore-forming bacterium are gram-positive rods of the genera *Bacillus* and *Clostridium*, although endospore-forming bacteria can be found in other genera.

These bacteria can be strict aerobes, facultative anaerobes, obligate anaerobes, or microaerophiles. *Microaerophiles* are bacteria that grow best in an environment that has a small amount of free oxygen.

The formation of endospores by bacteria is important in medicine and the food industry because these endospores are resistant to heat and many chemicals.

There are three major genera that produce endospores. These are

- *Bacillus.* *Bacillus* consist of the following bacterium:
  - *Bacillus anthracis.* *B. anthracis* causes *anthrax,* a severe blood infection of cattle, sheep, and horses that can be transmitted to humans. *B. anthracis* is a nonmotile facultative anaerobe that produces exotoxin. Anthrax can result in central nervous system distress, respiratory failure, anoxia, and death.
  - *Bacillus cereus.* *B. cereus* produces *enterotoxin* (a toxin that affects the intestine) and causes *gastroenteritis* (food poisoning).
  - *Bacillus thuringiensis.* *B. thuringiensis* produces a toxin that attacks the digestive system of insects, causing the insects to stop feeding by causing paralysis of the insect's gut. It is commonly used as a biological pesticide.
- *Sporosarcina.* *Sporosarcina* are bacteria that inhabit the soil and receive nutrients by feeding on dead organic matter.
- *Clostridium.* *Clostridium* are rod-shaped bacteria that exist in water, soil, and the intestinal tracts of animals and humans. These bacteria do not require oxygen. They release toxins that cause disease. Here are the common types of *Clostridium:*
  - *Clostridium tetani.* *C. tetani* causes *tetanus,* commonly referred to as *lockjaw.*
  - *Clostridium difficile.* *C. difficile* causes *gastroenteritis.*
  - *Clostridium perfringes.* *C. perfringes* causes *myonecrosis*—better known as *gas gangrene*—which produces hydrogen gas in deep tissue wounds, resulting in cell death.
  - *Clostridium botulinum.* *C. botulinum* is a cause of food poisoning, usually as a result of improperly canned food. It produces an exotoxin that causes flaccid paralysis (weakness of muscle tone) owing to the suppression of acetylcholine, which is a neurotransmitter. The result is vomiting, difficulty speaking, and difficulty swallowing, which can lead to respiratory paralysis and death. Physicians use toxin *C. botulinum* as a neural block that inhibits muscle contraction. *C. botilinum* is also used cosmetically to relax muscles that cause facial wrinkles (Botox injections). The *C. botulinum* toxin blocks the exocytosis of synaptic vesicles of the neuromuscular junction, where motor neurons meet muscle.

Gram-Positive Regular Non-Spore-Forming Bacilli

Regular non-spore-forming gram-positive rods are obligate or facultative anaerobes that have a rod-shaped appearance, as implied by the name. They inhabit fermenting plants and animal products. There are four genera within this section. These are

- *Lactobacillus.* Lactobacilli are nonsporing gram-positive rods. They are aerotolerant bacteria that produce lactic acid from simple carbohydrates. The acidity inhibits competing bacteria. An example of a *Lactobacillus* organism is the species *L. acidophilus.*
- *Lactobacillus acidophilus* is found in the human intestinal tract, oral cavity, and adult vagina. They produce an acidic environment that inhibits the growth of harmful bacteria by fermenting glycogen into lactic acid. *L. acidophilus* is also used commercially to produce an assortment of food products, including sauerkraut, pickles, buttermilk, and yogurt.
- *Listeria. Listeria monocytogenes* contaminate food and dairy products. If ingested, they can cause the disease *listeriosis,* which is an inflammation of the brain and meninges (meningitis).
- *Erysipelothrix. Erysipelothrixrhusiopathiae* causes erysipeloid, which are red, swollen, and painful lesions frequently seen in fishermen and meat handlers.

Gram-Positive Irregular Non-Spore-Forming Bacilli

These bacteria are irregular nonsporing rods. Although these bacteria generally are rod-shaped, they also can be pleomorphic. Some resemble a club, whereas others are long, threadlike cylinders. There are three genera within this section. These are

- *Corynebacterium.* Corynebacteria are club shaped and receive nutrients from dead or decaying organic material (*saprophytes*). Corynebacteria inhabit airy soil and water and cause diphtheria. *C. diphtheriae* is the organism that causes *diphtheria.*
- *Propionibacterium.* Propionibacteria infect wounds and cause abscesses. An example would be *P. acnes.*

- *Actinomyces*. *Actinomyces* are long, threadlike cylinders (filaments) that inhibit soil; some provide nitrogen to plants. The species *A. israelii* causes *actinomycosis*, which destroys tissues in the jaw, head, neck, and lungs. *Actinomyces* originally were classified as fungi because of their shape.

Mycobacteria

Mycobacteria require oxygen (aerobic) and are acid-fast organisms that remain red, whereas most are blue. Large numbers of lipids in the cell envelope of mycobacteria resist basic dyes. This organism got its name from the fact that *myco* means "fungus-like." There are three significant species:

- *Mycobacterium tuberculosis*. *M. tuberculosis* causes *tuberculosis*.
- *Mycobacterium leprae*. *M. leprae* causes *Hansen's disease* (leprosy).
- *Mycobacterium bovis*. *M. bovis* causes *tuberculosis* in cattle and can be transmitted to humans.

Nocardia

Nocardia are a group of a long, threadlike, cylinder-shaped bacteria that inhabit soil and need oxygen to grow (aerobic). They are gram-positive and cannot move by themselves (they are nonmotile).

N. asteroids is a common bacterium within this group. It causes *mycetoma*, which results in abscesses on the hands and feet and also causes lung infection.

QUIZ

Fill in the blanks:

1. This type of bacteria caused the black plague in Europe during the Middle Ages.
    _____.

2. When this organism reaches the lungs, it will produce pneumonia-like symptoms. This disease is now referred to as _____.

3. This bacteria can cause "rice water diarrhea." _____

4. This *Vibrio* type of bacteria can cause gastroenteritis when raw or undercooked fish is ingested. _____

5. This bacterium was identified by Louis Pasteur and is transmitted to humans by a dog bite. _____

Match the organism:

   A. *Shigella*
   B. *Salmonella enteritidis*
   C. *Neisseria mengitidis*
   D. *Haemophilus aegyptius*
   E. *Rickettsia rickettsii*

6. Causes inflammation of the meninges. _____

7. Causes shigellosis or traveler's diarrhea. _____

8. Causes salmonellosis, a form of food poisoning. _____

9. Causes conjunctivitis or "pink eye." _____

10. Causes Rocky Mountain spotted fever. _____

11

The Eukaryotes: Fungi, Algae, Protozoa, and Helminths

In this chapter, you will get close up and personal with eukaryotes. Eukaryotes are microorganisms commonly known as fungi, algae, protozoa, and helminthes and are in the kingdoms Fungi, Plants, Protists, and Animals.

CHAPTER OBJECTIVES

In this chapter, you will

- Learn about fungi
- Learn about algae
- Learn about protozoa
- Learn about helminths

Microorganisms can be beneficial to us because they help us to fight off pathogens and remove our waste, and they are a part of the food-making process such as that of yogurt. Yet microorganisms can interfere with our physiology, resulting in disease.

Fungi

Fungi have been studied systematically for 250 years, although ancient peoples learned of fermentation (enabled by fungi) thousands of years ago. Scientists who practice *mycology*, the study of fungi, are called *mycologists*. In the early days of microbiology, mycologists categorized fungi as plants because they resemble plants in general appearance (they have cell walls) and because both fungi and plants lack motility (neither can move under its own power). Fungi are categorized as either macroscopic, such as mushroom and puffballs, or microscopic, such as molds and yeast. Only molds and yeast care considered in microbiology.

Today, however, fungi and plants are considered two distinct groups of organisms because plants use chlorophyll to obtain nutrients, and fungi do not. Fungi are *heterotrophic:* They absorb nutrients from dead organic matter and organic waste (*saprophytes*) or tissues of other organisms (*parasites*). Many fungi are multicellular and are called *molds*. *Yeasts* are unicellular fungi.

Fungi can be both beneficial and harmful. For example, fungi called *mycorrhizae* are *mutualistic* and help the roots of plants absorb water and minerals from the soil. The cellulose and lignin of plants are important food sources for ants, but ants are unable to digest them unless fungi first break them down. Ants are known to cultivate fungi for this purpose. Some fungi are beneficial to humans as food (mushrooms). Fungi also are used in the preparation of food such as bread and beer (yeast). In addition, fungi are used to fight off bacterial diseases (antibiotics).

Some fungi can have a harmful effect because they feed on plants, animals, and humans, causing crop loss and destruction of animal tissue in their quest for nutrients. In humans, fungi cause various diseases such as athlete's foot.

Anatomy of Fungi

The body of a fungus (Fig. 11-1) is referred to as either a *soma* (meaning "body"), which is equivalent to the term *vegetative* in plants, or a *thallus*, which is also applied to algae and bryophytes (nonflowering plants comprised of mosses and liverworts). The body of a mold or fleshy fungus consists of long, loosely packed filaments called *hyphae*.

Hyphae are divided by cell walls called *septa* (the singular form is *septum*). In most molds, the hyphae are divided into segments by septa.

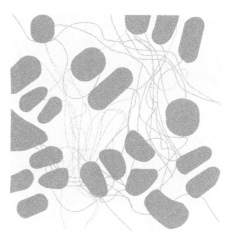

FIGURE 11-1 · The body of a fungus contains long filaments called hyphae.

In some fungi, the hyphae have no septa and look like long multinucleated cells called *coenocytic hyphae*. Cytoplasm flows or streams throughout the hyphae through pores in the septa. Under the right environmental conditions, the hyphae grow to form a *filamentous mass* known as a *mycelium*. A fungus can have a thallus many meters long that penetrates its surroundings. In the hyphae of fungi there is a portion called the *vegetative hyphae*. Vegetative hyphae are where nutrients are obtained. The part of the hyphae responsible for reproduction is called the *reproductive* or *aerial hyphae*.

Fungi can reproduce both sexually and asexually. Reproduction occurs with the formation of spores. Spores are always *nonmotile* and are a common means of reproduction among fungi. Do not confuse bacterial endospores with fungal spores; they are different. Bacterial endospores are formed so that the bacterial cell can survive in harsh environments. Once there is a less threatening environment, the bacterium leaves the endospore state and becomes active. The endospore germinates into a single bacterial cell. *Asexual reproduction* occurs when *asexual spores* are formed by the hyphae of a mold. When these spores germinate, they are identical to the parent. *Sexual reproduction* happens when the nuclei of *sexual spores* from two opposite mating strains of the same fungal species fuse. Fungi that grow from sexual spores have genetic characteristics of both parents.

Yeasts

Yeasts are fungi that are unicellular and reproduce using a process called *budding*. *Budding* occurs when a cell divides evenly to form two new cells. When the cell divides by *budding*, it divides unevenly. Yeasts are nonfilamentous and have a spherical or oval shape. The white powdery substance that is sometimes found coating fruits and leaves is a yeast. Yeasts also can reproduce sexually.

Molds

When a mold forms an asexual spore, the spore will detach itself from the parent and then germinate into a new organism. This process is considered reproduction because a second new organism grows from the spore.

Fungi Classification

Asexual fungal spores are formed on hyphae of fungi. When these hyphal spores germinate, they are identical to the parent. Asexual spores reproduce by the process of *cell division*. In sexual cell reproduction, the spores are produced by the fusion of nuclei from two opposite fungi of the same species. These fungi will have the characteristics of both parents. Asexual spores are produced more frequently than sexual spores. Asexual spores are present in virtually every environment on the planet.

Some fungi change their structure based on their habitat. This is referred to as *dimorphism*, the property of having two forms of growth. For example, some fungi appear nonfilamentous when growing outside their natural habitat but filamentous when growing in their natural habitat. Such changes in appearance can make it challenging to identify a particular type of fungus. Fungal classification is based on the type of sexual spores the fungi produce.

Listed below are examples of the divisions of the kingdom Fungi:

- *Zygomycota.* Zygomycota are conjugative fungi. They reproduce both sexually (zygospores) and asexually (sporangiospores). An example is *Rhizopus nigricans*, a black bread mold.
- *Ascomycota.* Ascomycota, also called *sac fungi*, have saclike cells called *asci*. These are yeasts, truffles, morels, and common molds. Fungi in this group reproduce sexually and asexually. Their sexual spores (*conidiosphores*) freely detach with the slightest movement (*conidia*) and therefore can cause infection (opportunistic disease) or an allergic reaction. Examples include
  - *Blastomyces. Blastomyces* cause *blastomycosis*, which is a general pulmonary disease.
  - *Histoplasma. Histoplasma* is a genus of fungus found in bird and bat droppings; it causes *histoplasmosis*, which is known as the *fungus flu*.

- *Basidiomycota*. Also called *club fungus*, basidiomycota include mushrooms, toadstools, the plant pathogen smuts, and rusts. Sexually produced *basidiospores* are formed externally on a base pedestal, producing a club-shaped structure called a *basidium* (plural *basidia*). Basidia can be found on visible fruiting bodies called *basidiocarps*, which are positioned on stalks. A mushroom is a basidiocarp. Some mushrooms, such as *Amanita*, produce toxins and are poisonous to humans, whereas others are very nutritious.

- *Deuteromycota*. Deutermycota, also known as *Fungi Imperfecti*, have no sexual reproduction (or none that can be observed). They cause *pneumocystis*, which infects people who have a compromised immune system.

Fungal Nutrition

Fungi receive nutrients by absorptive nutrition (*chemoheterotrophic*), which is somewhat similar to how bacteria obtain nutrients. Fungi team up with bacteria to break down organic molecules and are the principal decomposers on Earth. Fungi can metabolize complex carbohydrates, such as the lignin in wood.

Fungi can decompose substances that have very little moisture and substances that live in an environment with a pH of 5. Almost all molds are aerobic, and most yeasts are facultative anaerobes.

Still Struggling

Remember that yeasts and molds are *fungi*. Some examples include Zygomycota, Ascomycota, Basidiomycota, and Deuteromycotia. Types of *algae* include chrysophytes, diatoms, dinoflagellates, red algae, brown algae, and green algae. *Lichens* are filaments of fungi and cells of algae that live in a symbiotic relationship. *Protozoa*, organisms that are members of the kingdom Protista, include amoebas, flagellates, hemoflagellates, ciliates, and apicomplexans. *Flatworms* include flukes, tapeworms, round worms, and pinworms.

Algae

Algae are very simple unicellular or multicellular eukaryotic organisms that obtain energy from sunlight (*photoautotrophs*). They live in various water environments (oceans and ponds), on moist rocks and trees, and in soil.

Reproduction of Algae

Sexual reproduction occurs in most species of algae. In these species, the algae reproduce asexually for generations until there is a change in environmental conditions; then the algae reproduce sexually. Other types of algae alternate in how they reproduce. The algae that reproduce sexually will later reproduce asexually. All algae reproduce asexually. Unicellular algae divide by *mitosis* and *cytokinesis*. Multicellular algae that contain thalli and filaments can fragment. Each new piece can form a thallus and a filament.

Types of Algae

Chrysophytes

Chrysophyta can be divided into three groups: gold-brown, yellow-green, and diatoms (Fig. 11-2). *Chrysophytes* are unicellular algae that live in fresh water and contain *chlorophyll a* and *chlorophyll c*, which are *photosynthetic* pigments used to transform sunlight into energy. These are also known as *golden algae* because they have golden silica scales. There are 500 known species of chrysophytes. Some chrysophytes are amoeboid and attack bacteria by engulfing and destroying them.

Diatoms

Diatoms are unicellular algae that have a hard double outer shell made of silica. Nutrients pass through pores in the shell and then through the diatom's plasma membrane contained within the shell. There are 5,600 known species of diatoms, most of which are phototrophic and contain chlorophyll a and chlorophyll c pigments. They also contain *carotenoids*, which are yellow and orange pigments. Some diatoms are *heterotrophs* and break down and use organic matter as nutrients. Diatoms accumulate at the bottom of the sea and are mined commercially for both their value as an abrasive and their filtering and insulating capabilities (used in the filters in swimming pools).

Dinoflagellates

Dinoflagellates are unicellular algae that have the capability of self-movement through the use of tail-like projections called *flagella* (Fig. 11-3). The flagella are located between grooves in the cellulose plates that cover the dinoflagellate's body. These flagella pulsate in both an encircling motion around the body and in a perpendicular motion, causing the dinoflagellates to rotate like a top. There are about 1,200 known species of dinoflagellates that inhabit both fresh water and seawater.

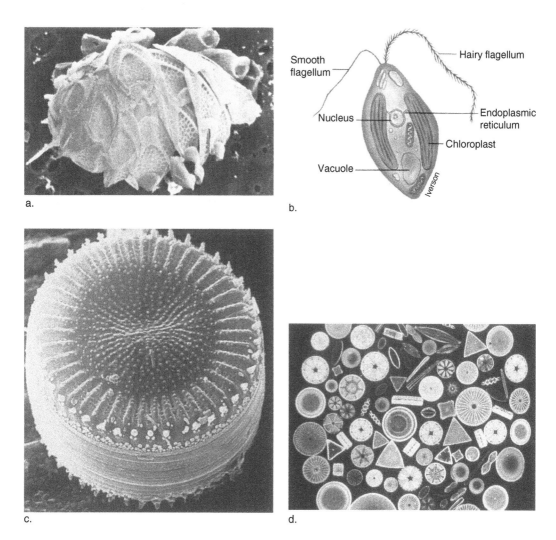

a.

b.

Smooth flagellum

Hairy flagellum

Nucleus

Endoplasmic reticulum

Chloroplast

Vacuole

Iverson

c.

d.

FIGURE 11-2 • *Chrysophyta* (yellow-green and golden-brown algae; diatoms). (a) Scanning electron micrograph of *Mallomonas*, a chrysophyte, showing its silica scales. The scales are embedded in the pectin wall but synthesized within the Golgi apparatus and transported to the cell surface in vesicles (×9,000). (b) *Ochromonas*, a unicellular chrysophyte. Diagram showing typical cell structure. (c) Scanning electron micrograph of a diatom, *Cyclotella meneghiniana* (×750). (d) Assorted diatoms as arranged by a light microscopist (×900). (From Prescott et al., *Microbiology,* 6th ed., McGraw-Hill, 1996.)

Dinoflagellates live in seawater. Some are heterotrophs and break down organic matter for nutrients. Some seawater dinoflagellates are luminous, giving a twinkle to the sea at night. Freshwater dinoflagellates are phototrophic: They synthesize nutrients from sunlight using photosynthesis.

Many dinoflagellates have chlorophyll a and c pigments as well as the yellow and orange carotenoid pigments. Depending on their photosynthetic pigment,

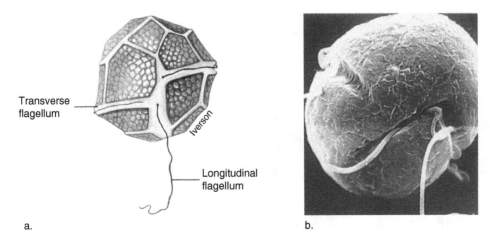

Transverse flagellum

Longitudinal flagellum

a.

b.

FIGURE 11-3 • Dinoflagellates. (a) *Ceratinum*. (b) Scanning electron micrograph of *Gymnodinium* (×4,000). Notice the plates of cellulose and the two flagella: one in the transverse groove and the other projecting outward. (From Prescott et al., *Microbiology,* 6th ed., McGraw-Hill, 1996.)

dinoflagellates can appear yellow-green, green, brown, blue, or red. When dino-flagellates undergo a population explosion, the sea changes color from blue to red or brown.

Some dinoflagellates, such as *Gonyaulax (plankton)*, can be fatal to humans because they produce neurotoxins. These dinoflagellates are eaten by fish and are absorbed by oysters, clams, and other shellfish (mollusks). The neurotoxins build up in their tissues, making the seafood poisonous to humans. These dino-flagellates also cause *red tides* that have a devastating effect on the fish population.

Red Algae

Red algae, also known as *rhodophytes*, are algae that form colonies in warm ocean currents and tropical seas (Fig. 11-4). They contribute to the formation of coral reefs that can be found as deep as 268 meters below the surface of the ocean and can grow to be 1 meter long. Their stonelike appearance is caused by a buildup of calcium carbonate deposits on their cell walls.

There are 4,000 known species of red algae, of which fewer than 100 are found in fresh water. Red algae get their color from the phycobilins and chlorophyll a pigments contained in their cells. *Phycobilin* pigment absorbs green, violet, and blue light, which are light waves that are capable of penetrating the deepest waters. It is for this reason that red algae can survive at great depths. The pigment that makes the algae red is called *phycoerythrin*.

FIGURE 11-4 · *Rhodophyta* (red algae). These algae (e.g., Corallina gracilis) are much smaller and more delicate than the brown algae. Most red algae have a filamentous, branched morphology as seen here. (From Prescott et al., *Microbiology*, 6th ed., McGraw-Hill, 1996.)

As you learned in Chapter 6, red algae are used to make agar. Agar is the culture medium that is extracted from the cell walls of red algae and is used to grow bacteria. Red algae are also used as a moisture-preserving agent in cosmetics and baked goods. Red algae also are used as a setting agent for jellies and desserts.

Brown Algae

Brown algae, also known as *phaeophytes*, are multicellular organisms (Fig. 11-5). Some brown algae are commonly called *kelp*; they live in the northern rocky shores of North America and can grow up to 30 meters in size. There are 1,500 known species of brown algae.

Brown algae have chlorophyll a and b photosynthetic pigments. They also have carotenoids. Brown algae can appear dark brown, olive green, and even golden depending on the types of pigments in their cells. The pigment that makes the algae brown is called *fucoxanthin*. *Algin* is a gummy substance found in the cell walls of some species of brown algae and is used as a thickening, emulsifying, and suspension agent for ice cream, pudding, frozen foods, toothpaste, floor polish, cough syrup, and even jelly beans.

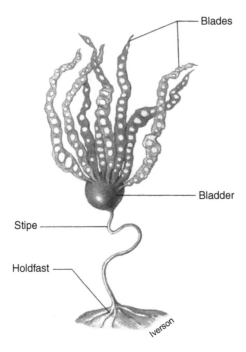

FIGURE 11-5 • *Phaeophyta* (brown algae). Diagram of the parts of the brown alga, Nereocystis. Due to the holdfast organ, the heaviest tidal action and surf seldom dislodge brown algae from their substratum. The stipe is a stalk that varies in length; the bladder is a gas-filled float. (From Prescott et al., *Microbiology*, 6th ed., McGraw-Hill, 1996.)

The organic matter that kelp produces supports the life of invertebrates, marine mammals, and fish.

Green Algae

Green algae can live in moist places on land, such as tree trunks, and in the soil, as well as in water. There are 7,000 species of green algae that are diverse in size, morphology, lifestyle, and habits. Scientists believe that some members of the species are linked structurally and biochemically to the Plant kingdom.

Two common green algae are

- *Spirogyra*. *Spirogyra* are freshwater algae that have tiny filaments, each containing spiraling bands of chloroplasts.
- *Volvox*. *Volvox* are colonial multicellular green algae that have flagella and live in marine, brackish, and freshwater environments.

Lichens

Lichens are filaments of a fungus and cells of algae (this is a *symbiotic* relationship) that are found on exposed soil or rock, on trees, on rooftops, and on cement structures. There are about 20,000 known species of lichens.

Survival of the green algae and the fungus is interdependent in a symbiotic association. Neither can live without the other. However, each grows independently. Lichens are delicate and beautiful in appearance.

Protozoa

These organisms are members of the kingdom Protista. There are about 20,000 known species of protozoa that live in water and soil. Some feed on bacteria, whereas others are parasites and feed off their hosts.

Most protozoa are asexual and reproduce in one of three ways. These are

- *Fission.* Fission occurs when a cell divides evenly to form two new cells.
- *Budding.* Budding occurs when a cell divides unevenly.
- *Multiple fission (schizogony).* Multiple fission occurs when the nucleus of the cell divides multiple times before the rest of the cell divides. A membrane forms around each nucleus when the nucleus divides, and then each nucleus separates into a daughter cell.

Some protists are sexual and exchange genetic material from one cell to another through *conjugation*, which is the physical contact between cells.

A protist can survive in an adverse environment by encapsulating itself with a protective coating called a *cyst*. The cyst defends the protist in extreme temperatures against toxic chemicals and even when there is a lack of oxygen, moisture, and food.

Protozoa Nutrition

Protists receive nutrients by breaking down organic matter (*heterotrophic*) and can grow in both aerobic and anaerobic environments, such as protists that live in the intestine of animals. Some protists, such as *Euglena*, receive nutrients from organic matter and through photosynthesis because they contain chlorophyll. These protists are considered both algae and protozoa.

Protists obtain food in one of three ways:

- *Absorption.* Food is absorbed across the protist's plasma membrane.
- *Ingestion.* Cilia outside the protist create a wavelike motion to move food into a mouthlike opening in the protist called a *cytostome.* An example is the paramecium.
- *Engulfment. Pseudopods* (meaning "false feet") on the protist engulf food and then pull it into the cell using a process called *phagocytosis.* An example of this type of protist is the amoeba.

Food is digested in the vacuole after the food enters the cell. The *vacuole* is a membrane-bound organelle. Waste products are excreted using a process called *exocytosis.*

Amoebas

Listed below are the common amoebas (Fig. 11-6):

- *Entamoeba histolytica.* This microorganism is transmitted between humans through the ingestion of cysts that are excreted in the feces of infected people. It is the causative agent of *amoebic dysentery* (Fig. 11-7).
- *Naegleria fowleri.* This amoeba causes *primary amoebic meningoencephalitis* (PAM), which results in headache, fever, vomiting, stiff neck, and loss of bodily control. *N. fowleri* enters the body through the mucous membranes (when the person swims in warm water) and travels to the brain and spinal cord.

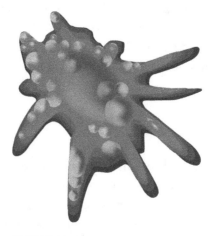

FIGURE 11-6 · A common type of amoeba is *Amoeba proteus.*

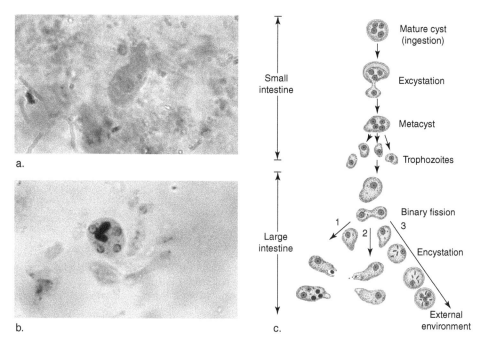

FIGURE 11-7 · *Amebiasis caused by Ent amoeba histolytica.* (a) Light micrographs of a trophozoite (×1,000) and (b) a cyst (×1,000). (c) Life cycle. Infection occurs by the ingestion of a mature cyst of the parasite. Excystation occurs in the lower region of the small intestine and the metacyst rapidly divides to give rise to eight small trophozoites (only four are shown here). These enter the large intestine, undergo binary fission, and may (1) invade the host tissues, (2) live in the lumen of the large intestine without invasion, or (3) undergo encystation and pass out of the host in the feces. (From Prescott et al., *Microbiology*, 6th ed., McGraw-Hill, 1996.)

- *Acanthamoeba polyphaga.* This amoeba infects the cornea of the eye and causes *keratitis* (inflammation of the cornea), leading to blindness. It also can cause ulcerations of the eye and the skin. *A. polyphaga* is also known to invade the central nervous system, resulting in death.

Amoebas live in fresh water and soil that is moist.

Flagellates

Flagellates move by structures called *flagella*. They have two or more spindle-shaped flagella in the front of the cell that they use to pull themselves through their environment. Food enters flagellates through a mouthlike groove called a *cytostome*.

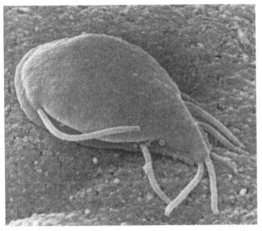
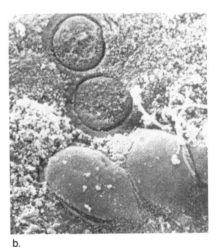

a.   b.

FIGURE 11-8 · Giardiasis. (a) *Giardia lamblia* adhering to the epithelium by its sucking disk; scanning electron micrograph. (b) Upon detachment from the epithelium, the protozoa often leave clear impressions on the microvillus surface (upper circles); scanning electron micrograph. (From Prescott et al., *Microbiology*, 6th ed., McGraw-Hill, 1996.)

Here are some common flagellates:

- *Trichomonas vaginalis.* This flagellate is the cause of *trichomoniasis*, which is a sexually transmitted disease. *T. vaginalis* is found in the male urinary tract and the vaginas of females.

- *Giardia lamblia.* This flagellate causes *giardiasis* (Fig. 11-8). Giardiasis results in nausea, cramping, and diarrhea when food or water contaminated by fecal material is ingested. *G. lamblia* lives in the small intestines of humans and other mammals.

Blood and Tissue Protozoa

Hemoflagellates are protozoa that are carried by blood-feeding insects and are transmitted into the bloodstream by the insect's bite. Here are some commonly found hemoflagellates:

- *Trypanosoma gambiense.* *T. gambiense* is transmitted in the saliva of the tsetse fly as a result of a bite and causes *trypanosomiasis*, which is better known as *African sleeping sickness*.

- *Trypanosoma cruzi.* *T. cruzi* is carried by the reduviid bug, which is commonly called the *assassin bug* because it bites the face. *T. cruzi* causes

Chagas disease. Chagas disease is thought to have made Charles Darwin sick during his voyage on the *H.M.S. Beagle*. Typically, there aren't any symptoms for months after the bite. During this time, *T. cruzi* spreads throughout the organs of the body, weakening the heart, intestines, and esophagus. It also causes both eyes to swell (*Romaña's sign*).

Ciliates

Ciliates are protozoa that have shorter, hairlike structures called *cilia* that are found in rows on the outer surface of the cell (Fig. 11-9). These cilia are used to move the protozoa through the environment and are used to bring food into the cell.

An example of a ciliate is *Balantidium coli*. This is the only ciliate that causes disease in humans. When ingested, it enters the large intestine, causing severe dysentery.

Apicomplexans

Apicomplexans are sporozoa protozoa that can live and grow inside another living organism. They are either intracellular or extracellular parasites of animals. They cannot move by themselves (nonmotile). Apicomplexans have an apical complex of organelles that forms an apex (tip). This apex contains enzymes that enable it to penetrate the tissues of a host. Apicomplexans also go through a spore-forming stage.

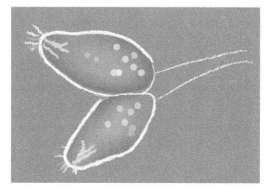

FIGURE 11-9 • Ciliates are protozoa that have cilia.

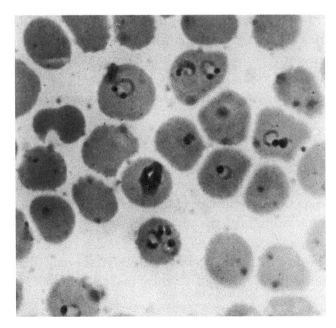

FIGURE 11-10 • Malaria: Erythrocytic Cycle. Trophozoites of *P. falciparum* in circulating erythrocytes; light micrograph (×1,100). The young trophozoites resemble small rings resulting in the erythrocyte cytoplasm. (From Prescott et al., *Microbiology*, 6th ed., McGraw-Hill, 1996.)

Here are common apicomplexans:

- *Plasmodium. Plasmodium* is an important sporozoa parasite. It is the organism that causes *malaria*, which means "bad air" (Fig. 11-10). *Plasmodium* lives in the female *Anopheles mosquito* and causes malaria when the mosquito bites a human. Symptoms of malaria include severe chills and fever or *rigor* (a sudden chill or coldness that is followed by fever. *There are four species of Plasmodium that infect humans and cause malaria: P. falciparum, P. malariae, P. vivax,* and *P. ovale.*

  When the female mosquito feeds on her victim, saliva containing the sporozoites gets injected into the bloodstream of the victim and travels to the cells of the liver. In the liver cells, the sporozoites undergo multiple asexual fission and make *merozoites,* which invade red blood cells. In the red blood cells, the organism grows into a cell called a *trophozoite.* The red blood cells will break apart (*lysis*), releasing more merozoites into the blood and thus infecting more red blood cells.

- *Bebesia microti*. *B. microti* lives in ticks and causes *bebesiosis* when the tick bites someone. *B. microti* then enters red blood cells, where it multiplies quickly. At first, there aren't any symptoms (*asymptomatic*). However, soon afterwards, there is a high fever, headache, and muscle pain as *B. microti* destroys red blood cells. This causes the person to become *anemic* (insufficient hemoglobin because of the reduction in the number of red blood cells) and show signs of *jaundice* (an increase in bile, causing the skin and eye sclera—the white part of the eye—to yellow).

- *Toxoplasma gondii*. *T. gondii* can be found in most birds and practically all animals, although cats are a definite host. *T. gondii* lives in cat feces and raw meat and causes *lymphadenitis* (infection of the lymph nodes), which can have a devastating effect on people who are immune compromised, such as patients with AIDS. *T. gondii* causes congenital infections in a fetus because it can pass from the mother to the fetus through the placenta.

Helminths

Helminths are parasitic multicellular eukaryotic animals. Most of these animals belong to the phyla Platyhelminthes and Nematoda. There are free-living members of these phyla, but in this section, only the parasitic organisms will be discussed.

Many parasitic helminths do not have a digestive system and instead absorb nutrients from the food that is consumed by their host organism, the host's body fluids, and the host's tissues. Parasitic helminths have a very simplistic nervous system because they have to respond to very few changes in their host's environment. They lack or have reduced means of locomotion because they are transferred from one host to another. Parasitic helminths have a very complex reproductive system that produces fertilized eggs (*zygotes*) that infect the host organism.

Life Cycle of Helminths

The life cycle of parasitic helminths that go through a developmental larval stage involves an intermediate host. *Dioecious* adult helminths are male or female. That is, one organism has a male reproductive system, and another has a female reproductive system. When these two adult helminths with different sex organs occupy the same host organism, sexual reproduction can occur.

Monoecious adult helminths are hermaphroditic (an organism that has both female and male reproductive organs). Some of these monoecious helminths can fertilize themselves, whereas others may fertilize each other.

Flatworms

Flatworms, also known as *platyhelminths*, are mostly parasitic aquatic organisms that range in size from 1 millimeter to 10 meters, as in the case of a tapeworm. There are more than 15,000 known species of flatworms. A flatworm has both male and female reproductive parts (*monoecious*). Most, but not all, of their oxygen and nutrients is absorbed through their body wall.

There are two categories of flatworms:

- *Flukes.* Flukes are flat, leaf-shaped bodies that have an oral and a ventral sucker that are used to hang onto the body of a host. Flukes live inside the intestines or on tissues of humans. Three common flukes are
  - *Schistosoma.* This genus of flukes causes the disease *schistosomiasis*, a debilitating disease that causes portal hypertension and liver cirrhosis.
  - *Paragonimus westermani.* This fluke causes *paragonimiasis*, which is the result of the fluke's depositing eggs into the bronchi of the lungs.
  - *Clonorchis sinensis.* Also known as *Chinese liver fluke*, this fluke causes *clonorchiasis*, which occurs when the fluke latches itself inside the liver.
- *Tapeworms.* Tapeworms have a knoblike head, called a *scolex*, with hooks that allow it to attach to the walls of the intestines of vertebrate animals (including humans). Tapeworms have a series of flat, rectangular body units called *proglottids* (compartments that contain reproductive organs). Proglottids eventually break away from the tapeworm and are excreted in feces. However, new proglottids take their place. A tapeworm continues to grow as long as its scolex and neck are intact. Common tapeworms include
  - *Taenia saginata.* Also called *beef tapeworm*, this tapeworm lives in raw or poorly cooked beef and can cause *taeniasis*. *T. saginata* can grow to a length of 25 meters in the intestines of a human, leading to an intestinal blockage and malnutrition as the tapeworm absorbs nutrients intended for the person.
  - *Tania solium.* Also called the *pork tapeworm*, this tapeworm lives in raw or poorly cooked pork and can cause *taeniasis*. *T. solium* can grow to a length of 7 meters in the intestine, leading to an intestinal blockage and malnutrition.

- *Echinococcus granulosus.* This tapeworm is spread to humans through contact with an infected dog and is transmitted when a dog licks a person. This can lead to infection, anaphylactic shock, and death if the tapeworm enters the body. *E. granulosus* can lay eggs that produce cysts called *hydatid cysts* in the lungs, liver, and brain.
- *Hymenolepis nana.* This is a tapeworm that lays eggs in cereals and foods that are contaminated with infected parts of insects. When someone ingests the cereal or food, he or she also ingests the tapeworm. The tapeworm then attaches to the intestines, leading to diarrhea, abdominal pain, and convulsions.
- *Diphyllobothrium latum.* This is a broad fish tapeworm that lives in raw or poorly cooked fish. The tapeworm attaches to the intestines of the fish, where it then lays eggs. While a person ingests the fish, the tapeworm attaches to the human intestine and absorbs large quantities of vitamin B_{12} from the intestine, eventually causing the person to develop *vitamin deficiency anemia.* This is also called *pernicious anemia* because there is insufficient vitamin B_{12} to make red blood cells.

Roundworms (Nematodes)

Roundworms are also known as *nematodes* and live in soil, fresh water, and saltwater. Most of the over 80,000 species of roundworms are parasites and live in plants or animals such as insects. They have a primitive body that consists of a cylindrical tube that has tapered ends and is covered with a thick protective layer called a *cuticle*.

Common roundworms include

- *Ascaris lumbricoides.* This is a roundworm that is transmitted by contaminated human fertilizer, food, or water. It causes *ascariasis*, which is an infection of the small intestine.
- *Strongyloides stercoralis.* This is a roundworm whose larvae penetrate human skin and spread into the small intestine, where they cause *strongyloidiasis*, which is an infection of the small intestine.
- *Trichinella spiralis.* This is a roundworm whose larvae live in undercooked meats, mainly pork, and cause *trichinosis*. These juvenile worms in ingested meat mature in the small intestine of the host organism. The mature females burrow through the wall of the small intestine and release their offspring (juveniles) into the blood of the host; skeletal muscle infection soon follows

as these juveniles burrow into the skeletal muscle of the host, where they form into a sack (*encyst*) causing muscle pain and fever; this results in a large number of eosinophilic leukocytes (*eosinophilia*). An *eosinophilic leukocyte* is a type of white blood cell that increases with allergies and infections.

- *Wuchereria bancrofti.* This is a roundworm that lives in mosquitoes and causes *elephantiasis* when the infected mosquito bites a human. The mosquito injects its larvae into the skin, and the larvae then migrate to the lymph nodes, causing blockages.

- *Onchocerca volvulus.* This is a roundworm that lives in the blackfly and causes *river blindness* when the blackfly bites a human.

- *Dracunculus medinensis.* This is a roundworm that lives in lobsters, crabs, shrimps, and other Crustacea. When the infected organism is ingested, this roundworm's larvae migrate from the person's intestines through the abdominal cavity to subcutaneous tissue, where they mature. *D. medinensis* releases a toxic substance that creates a skin ulcer, which is the common symptom of *dracunculosis disease.*

- *Hookworms* are roundworms that have tiny hooks that are used to attach to a host, typically in the intestine. Here are some common hookworms:
  - *Necator americanus.* Also known as the *New World hookworm*, this hookworm lives in the lower intestine and is the second most common hookworm infection. Its eggs are passed into the feces. Once it comes into contact with a human, it penetrates the skin and spreads into the heart, lungs, and eventually, the small intestine, where it grows into an adult. This can lead to severe blood loss and anemia.
  - *Ancylostoma duodenale.* Also known as the *Old World hookworm*, this hookworm is similar to *N. americanus* but is native to southern Europe, North Africa, northern Asia, and parts of western South America.
  - *Ancylostoma braziliense.* This is a hookworm that exists in cats and dogs and causes *cutaneous larva migrans*, which is also known as *creeping eruption*. Its eggs are passed in the feces, and the larvae develop in the soil. The larvae can tunnel into the epidermis of humans and can cause an infection to develop.
  - *Ancylostoma caninum.* This is a hookworm that exists in dogs and frequently infects puppies. It eats away at the tissues in the small intestine and sucks blood from the dog. This can result in diarrhea, weight loss, anemia, and death. The larvae can tunnel into the epidermis of humans and can cause an infection to develop.

Pinworms

Pinworms can be up to 10 millimeters long and live in the large intestine. Female pinworms crawl out the anus to lay eggs on the perianal skin. Afterwards, the female dies. Pinworms infect about 10 percent of humans, although a person may not know that he or she is infected. Pinworms cause few or no symptoms besides a mild gastrointestinal upset and perianal itching, which can lead to bacterial infections. Pinworms are highly contagious and can be transmitted in bed linens and clothing that has been contaminated with eggs. A common pinworm is *Enterobius vermicularis.* It is the most common pinworm in the United States. *E. vermicularis* causes *enterobiasis*, in which the skin around the anus is so itchy that a person might not be able to sleep.

QUIZ

Fill in the blanks:

1. A fungus that causes black bread mold. _____

2. A type of fungus found in bird and bat droppings. _____

3. This type of algae is also known as *golden algae*. _____

4. This algae can move itself with flagella. _____

5. This type of dinoflagellate can cause the red tide. _____

Match the parasite:

   A. *Trypanosoma gambiense*
   B. *Trichomonas vaginalis*
   C. *Entamoeba histolytica*
   D. *Trypanosoma cruzi*
   E. *Toxoplasma gondii*

6. Causes amoebic dysentery. _____

7. Causes trichomoniasis. _____

8. Causes African sleeping sickness. _____

9. Causes Chagas disease. _____

10. Lives in cat feces. _____

Viruses, Viroids, and Prions

In this chapter, you'll learn about viruses—how they work and viral diseases. You'll also learn about viroids and prions, which are molecules that also cause infection.

CHAPTER OBJECTIVES

In this chapter, you will

- Learn about viruses
- Become familiar with the lysogenic cycle
- Learn about animal viruses
- Learn about viroids
- Learn about prions

Recall the time when your nose was running and you had a hacking cough. No doubt you went to the doctor and asked for antibiotics to cure your head cold. Surprise! You were not prescribed an antibiotic. The doctor probably said that you didn't have a bacterial infection. You had a viral infection. And you were sent home with a combination of over-the-counter medication and home remedies and told to let the virus run its course.

Viruses are tiny microorganisms that take over your own cells, making it difficult for your immune system to kill them because the infected cells look to your immune system like all your other cells.

Viruses

In 1889, Dutch plant microbiologist Martinus Beijerinck described the concept of viruses through his studies of tobacco mosaic disease. Nobel prize winner Sir Peter Medawar described how microbiologists feel about viruses when he said, "A virus is a piece of bad news wrapped in a protein."

Viruses are strands of nucleic acid that are encased within a protein coat, making them difficult to destroy. A microorganism needs both DNA and RNA to reproduce. A virus cannot express genes without a host because a virus has either DNA or RNA but not both.

A *virus* (Latin for "poison") is an obligate intracellular parasite having only one type of nucleic acid, DNA or RNA, and it can replicate only inside a living host cell. Once inside the living host cell, a virus becomes integrated in the metabolism of its host, making a virus difficult to control by chemical means. You cannot kill a virus with antibiotics. Drugs that destroy the host's ability to be used by a virus for replication tend also to be highly toxic and have a negative and sometimes deadly effect on the host cell.

Before a virus enters a cell, it is a free virus particle called a *virion*. A virion cannot grow or carry out any biosynthetic or biochemical function because it is metabolically inert. Viruses are not cells. They vary in size from 20 nanometers (poliovirus) to 300 nanometers (smallpox virus) and cannot be seen under a light microscope.

In 1933, microbiologist Wendell Stanley of the Rockefeller Institute for Medical Research showed that viruses could be regarded as chemical matter rather than as living organisms.

Viral Structure

The major components of a virus are

- *Nucleic acid core.* The nucleic acid core can be either DNA or RNA that makes up the genetic information (genome) of the virus. RNA genomes occur only in viruses.
- *Capsid.* The capsid is the protein coat that encapsulates a virus and protects the nucleic acid from the environment. It also plays a role in how some viruses attach to a host cell. A capsid consists of one or more proteins that are unique to the virus and that determine the shape of the virus.
- *Envelope.* The envelope is the membrane bilayer that some viruses have outside their capsid. If a virus does not have an envelope, the virus is called a *naked virus.* A naked virus is more resistant to change and is less likely to be affected by conditions that can damage the envelope. Environmental factors that can damage the envelope include
  - Increased temperature
  - Freezing temperature
  - pH below 6 or above 8
  - Lipid solvents
  - Some chemical disinfectants (e.g., chlorine, hydrogen peroxide, and phenol)

Naked viruses are more resistant to changes in temperature and pH. Examples of diseases caused by naked viruses include poliomyelitis, warts, and the common cold.

Shapes of Viruses

A virus can have one of two structures. These are

- *Helical virus.* A helical virus is rod- or thread-shaped. The virus that causes rabies is a helical virus.
- *Icosahedral virus.* An icosahedral virus is spherically shaped. Viruses that cause poliomyelitis and herpes simplex are icosahedral viruses.

How Viruses Replicate

The easiest way to understand how viruses replicate is to study the life cycles of viruses called *bacteriophages.* Bacteriophages are viruses that infect bacteria and replicate by either a lytic cycle or a lysogenic cycle. The difference between these two cycles is that the cell dies at the end of the lytic cycle and remains alive at the end of the lysogenic cycle.

The first two scientists to observe bacteriophages were Frederic Twort of England and Felix d'Herelle of France in the early 1900s. The name *bacteriophage* is credited to d'Herelle and means "eaters of bacteria."

Lytic Cycle

The most studied bacteriophage is the *T-even bacteriophage*, which infects *Escherichia coli*. The virions of T-even bacteriophages are large and complex and do not contain envelopes. T-even bacteriophages are composed of a head and tail structure and contain genomes of double-stranded DNA. The *lytic cycle* of replication begins with the collision of the bacteriophage and a bacterium, called *attachment*. The tail of the bacteriophage attaches to a receptor site on the bacterial cell wall. After attachment, the bacteriophage uses its tail like a hypodermic needle to inject its DNA (nucleic acid) into the bacterium. This is called *penetration*. The bacteriophage uses an enzyme called *lysozyme* in its tail to break down the bacterial cell wall, enabling it to inject its DNA into the cell. The head or capsid of the bacteriophage remains on the outside of the cell wall. After the DNA is injected into the host's bacterial cell's cytoplasm, *biosynthesis* occurs. Here, the T-even bacteriophage uses the host bacterium's nucleotides and some enzymes to make copies of the bacteriophage's DNA. This DNA is transcribed to messenger RNA (mRNA), which directs the synthesis of viral enzymes and capsid proteins. Several of these viral enzymes catalyze reactions that make copies of bacteriophage DNA. The bacteriophage DNA then will direct the synthesis of viral components by the host cell.

Next, *maturation* occurs. Here, the T-even bacteriophage DNA and capsids are put together in order to make *virions*.

The last stage is the *release* of virions from the host bacterial cell. The bacteriophage enzyme lysozyme breaks apart the bacterial cell wall (*lysis*), and the new virus escapes. The escape of this new bacteriophage virus allows the new bacteriophage virus to infect neighboring cells, and the cycle will continue in these cells.

The Lysogenic Cycle

Some viruses do not cause lysis and ultimate destruction of the host cells they infect. These viruses are called *lysogenic phages* or *temperate phages*. These bacteriophages establish a stable, long-term relationship with their host called *lysogeny*. The bacterial cells infected by these phages are called *lysogenic cells*.

The most studied bacteriophage that multiplies using the lysogenic cycle is the bacteriophage *lambda*. This bacteriophage infects the bacterium *E. coli*.

When the bacteriophage lambda penetrates an *E. coli* bacterium, the bacteriophage DNA forms a circle. It recombines with the circular DNA of the bacterium. This bacteriophage DNA is called a *prophage* and has a linear piece of DNA that integrates directly into the host chromosome. The combination of viral and bacterial DNA is called prophage.

Every time the bacterial host cell replicates normally, so does the prophase DNA. On occasion, however, the bacteriophage DNA can break out of the prophage and initiate the lytic cycle.

Animal Viruses

Animal viruses infect and replicate animal cells. They differ from bacteriophages in the way they enter the host cell. For example, DNA viruses enter an animal host in this way:

- *Attachment*. Animal viruses attach to the host cell's plasma membrane proteins and glycoproteins (host cell receptors).
- *Penetration*. Animal viruses do not inject nucleic acid into the host eukaryotic cell. Instead, penetration occurs by *endocytosis*, where the virion attaches to the microvillus of the plasma membrane of the host cell. The host cell then enfolds and pulls the virion into the plasma membrane, forming a vesicle within the cell's cytoplasm.
- *Transcription* in the nucleus by host RNA polymerase.
- *Translation* by host cell ribosomes.
- *DNA replication* by host DNA polymerase in the nucleus.
- *Assembly* of viral particles.
- *Release* of viral particles from the cell by lysis or exocytosis.

RNA Viruses

The nucleic acid of a virus can be composed of either DNA (double-stranded or single stranded) or RNA (double-stranded or single-stranded). A virus does not possess both DNA and RNA.

- Attachment to host cell receptor
- Fusion with membrane of host

- Entry of the nucleocapsid into the cytoplasm
- Transcription in the cytoplasm by viral RNA polymerase
- Translation by host cell ribosomes
- Assembly of viral particles
- Release from cell

Viruses and Infectious Disease

Viruses are classified by the type of nucleic acid they contain, chemical and physical properties, shape, structure, host range, and how they replicate.

DNA Viruses

DNA viruses are viruses that have DNA but no RNA. Common DNA viruses include

- *Hepadnaviruses.* Hepadnaviruses cause serum hepatitis. The hepatitis B virus (HBV) is a common form of this virus that enters the body via hypodermic needles, blood transfusion, or sexual relations. (Hepatitis A, C, D, E, F, and G viruses are not related and are RNA viruses.)
- *Herpesviridae.* Herpesviruses (Fig. 12-1) cause the herpesvirus infection. There are about 100 forms of herpesviruses, including
  - *Herpes simplex virus type 1 (HSV-1).* Herpes simplex virus type 1 causes encephalitis and enters the body through lesions on the lip, skin, or eyes.
  - *Herpes simplex virus type 2 (HSV-2).* Herpes simplex virus type 2 is sexually transmitted, affects the genital and lip areas, and can lead to carcinomas.
  - *Varicella-zoster virus (VZV).* Varicella-zoster virus causes chickenpox in the acute form and shingles in the latent form. Shingles appear as vesicles along a nerve, resulting in severe pain along the course of the nerve.
  - *Cytomegalovirus (CMV).* Cytomegalovirus causes an infection that usually goes unnoticed unless the person's immune system is compromised, such as in the bodies of AIDS patients or in infants whose immune systems are not fully developed. This virus can be fatal to some infants.
  - *Epstein-Barr virus (EBV).* Epstein-Barr virus causes infectious mononucleosis.

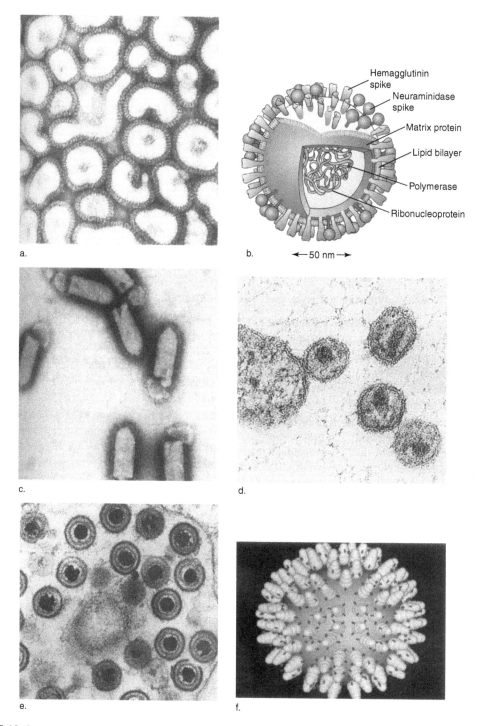

FIGURE 12-1 · Examples of Enveloped Viruses. (a) Human influenza virus. Note the flexibility of the envelope and the spikes projecting from its surface (×282,000). (b) Diagram of the influenza virion. (c) Rhabdovirus particles (×250,000). This is the vesicular stomatitis virus, a relative of the rabies virus, which is similar in appearance. (d) Human immunodeficiency viruses (×33,000). (e) Herpesviruses (×100,000). (f) Computer image of Semliki Forest virus, a virus that occasionally causes encephalitis in humans. (From Prescott et al., *Microbiology,* 6th ed., McGraw-Hill, 1996.)

- *Papovaviridae.* Papovaviruses, such as the human papilloma virus (HPV), cause warts (papillomas) and has been linked to cervical cancer in women.
- *Poxviridae.* Poxviruses cause pox (pus-filled lesions) diseases such as smallpox.
- *Adenoviridae.* Adenoviruses cause acute respiratory disease. This is the common cold virus.

RNA Viruses

An RNA virus is a virus that contains RNA but not DNA. Common RNA viruses include

- *Flaviviridae.* Flaviviruses, such as the dengue virus that causes break bone fever, are carried by mosquitoes. Break bone fever results in skin lesions, fever, and muscle and joint pain and is often fatal. Other flaviviruses include
  - *St. Louis encephalitis virus (Yellow Fever).* St. Louis encephalitis virus causes an infection that is not easily recognized. Wild birds and mosquitoes carry St. Louis encephalitis virus. Monkeys carry a form of this virus called *yellow fever virus,* which is transmitted to humans by mosquitoes and leads to severe liver damage.
  - *Hepatitis C virus (HCV).* Hepatitis C virus is called *non-A, non-B hepatitis virus* and results in chronic infection. Humans contract this virus from needle pricks and blood transfusions.
- *Picornavirus.* Picornaviruses such as the *poliovirus* cause *poliomyelitis* and kill motor neurons, resulting in weakness and loss of muscle tone (*flaccid paralysis*). Others include
  - *Hepatitis A virus (HAV).* Hepatitis A virus is also known as *infectious hepatitis* and is transmitted through a fecal-oral route.
  - *Rhinovirus.* This is the most frequent cause of the common cold. It causes localized upper respiratory tract infections.
- *Retroviridae.* Retroviruses are a group of RNA viruses that include the following commonly recognized viruses:
  - *Human immunodeficiency virus (HIV).* Human immunodeficiency virus (Fig. 12-1) (lentivirus) is a virus that often results in acquired immune deficiency syndrome (AIDS). This virus attacks T cells. A *T cell* is a white blood cell that fights infection and kills spontaneously arising tumors. HIV causes *Kaposi's sarcoma,* which is a rare form of cancer, and *Pneumocystis carinii* infection, which is an opportunistic infection that results in pneumonia in AIDS patients.

- *Human T cell leukemia viruses 1 and 2.* Human T cell leukemia viruses 1 and 2 are the viruses that causes acute *T cell lymphocytic leukemia* and often contain genes that cause cancer (*oncogenic*).

- *Togaviruses.* Togaviruses are viruses such as *Estera equine encephalitis* that are transmitted mainly through blood-sucking insects (*arbovirus*), such as mosquitoes. They cause severe encephalitis. Another is the *rubella virus*. The rubella virus causes *German measles*, which can be very dangerous if contracted during the first 10 weeks of pregnancy. The *rubella vaccine* is used to weaken the disease-producing ability of the rubella virus.

- *Orthomyxoviruses.* Orthomyxoviruses, such as *influenza viruses A, B, and* C, cause localized infection of the respiratory tract, which is usually not serious unless the infected person is elderly or the person is infected with secondary bacterial pneumonia. Influenza viruses A and B can cause *Guillain-Barré syndrome*, which is an inflammation of the nerves that are outside the brain and spinal cord (peripheral nerves); it appears three to five weeks after a person contracts the flu or after the person receives a flu vaccine. Influenza virus B causes *Reyes syndrome*, which is lethal to the liver and the brain and causes a brain disease (encephalopathy) following a mild flu, chickenpox, or the administration of aspirin.

- *Paramyxovirus.* Paramyxoviruses such as the *parainfluenza virus (Sendai virus)* cause croup in infants. Two other types of paramyxoviruses include
  - *Mumps virus.* The mumps virus causes an enlargement of one or both parotid glands and swelling and pain in the testes and ovaries. There is a vaccine to protect humans from the mumps virus.
  - *Measles virus.* The measles virus, which is also known as *rubeola*, causes measles. The measles virus causes a slow degeneration of the nervous system of teenagers and young adults. If not treated, measles can progress into encephalomyelitis or pneumonia.

- *Rhabdovirus.* Rhabdoviruses (Fig. 12-1) such as the *rabies virus (lyssavirus)* cause rabies following an animal bite. In rare cases, a person can be infected by inhaling the virus. Some animals such as bats pass the rabies virus through their feces.

- *Filoviridae.* An example of a filovirus would be the *ebola virus*, which causes African hemorrhagic fever.

Oncogenic Viruses

Oncogenic viruses are viruses that produce tumors when they infect humans. The more common oncogenic viruses include

- *Human papillomavirus (HPV)*. Human papillomavirus causes *common warts* but also is believed to cause *cervical cancer*.

- *Epstein-Barr virus (EBV)*. Epstein-Barr virus causes *Burkitt's lymphoma*, which is a tumor of the jaw. It is seen mainly in African children and causes a tumor in the nasopharyngeal (*nasopharyngeal carcinoma*).

- *Herpes simplex virus type 2 (HSV-2)*. Herpes simplex virus type 2 causes *genital herpes*, *cervical cancer* (cervical carcinoma), and *oral lesions*.

- *Human T cell leukemia virus 1 (HTLV-1)*. Human T cell leukemia virus 1 causes acute *T-cell lymphocytic leukemia*, which is a cancer that affects T-cell-forming tissues.

- *Human T cell leukemia virus 2 (HTLV-2)*. Human T cell leukemia virus 2 causes *atypical hairy cell leukemia*.

Still Struggling

Here is a quick guide to DNA and RNA viruses:

DNA viruses: Hepatitis B, Herpes simplex viruses types 1 and 2, Varicella-zoster (chickenpox), Cytomegalovirus, Epstein-Barr virus (mononucleosis), Human papillomavirus (warts and tumors), Adenovirus (common cold virus).

RNA viruses: Dengue virus (break bone virus), Hepatitis C virus (non-A, non-B hepatitis, Poliovirus (poliomyelitis), Hepatitis A virus (fecal-oral route of transmission), Rhinovirus (common upper respiratory tract infection).

Retroviruses (a group of RNA viruses): Human immunodeficiency virus (HIV), Togaviruse (Eastern equine encephalitis), Rubella (German measles), Orthomytovirus (influenza A, B, and C), Rubeola (measles), Mumps (mumps), Rhabdovirus (rabies).

Oncogenic viruses (viruses that can produce tumors): Human papillomavirus, Herpes simplex virus type 2, Human T cell leukemia virus 1, Human T cell leukemia virus 2.

Plant Viruses

Some viruses cause diseases in plants.

RNA plant viruses include

- *Picornaviridae*. These include the *bean mosaic virus*.
- *Reoviruses*. These include the *wound tumor virus*.

DNA plant viruses include the

- *Papovaviridae*, such as the *cauliflower mosaic virus*.

Viroids

In 1971, plant pathologist O. T. Diener discovered an infectious RNA particle that was smaller than a virus that causes diseases in plants. He called it a *viroid*. Viroids infect potatoes, causing potato spindle-tuber disease. They also can infect chrysanthemums by stunting their growth and can cause cucumber pale fruit disease. Millions of dollars are lost each year in crop failures caused by viroids.

A viroid is similar to a virus in that it can reproduce only inside a host cell as particles of RNA. However, it differs from a virus in that each RNA particle contains a round or circular single-stranded RNA molecule. As in some viruses, a viroid does not have a capsid or an envelope.

Prions

A *prion* is a small infectious particle that contains a component of protein. Some researchers believe that a prion consists of a protein without nucleic acid because a prion is too small to contain nucleic acid and because a prion is not destroyed by agents that digest nucleic acids.

Prion diseases, referred to as *transmissible spongiform encephalopathies* (TSEs), are progressive neurologic diseases that are fatal to humans and animals. Researchers believe that prions cause *Creutzfeldt-Jakob disease*, which is a neurologic disease that results in progressive dementia first observed by Hans Gerhard Creutzfeldt and Alfon Maria Jakob in the 1920s. In 1976, Carlton Gajdusek won the Nobel prize for his work with the TSE *kuru*. Kuru is characterized by progressive ataxia, incapacitation, and death.

In 1982, neurobiologist Stanley Prusiner proposed that proteins cause the neurologic disease *scrapie*, which is a degenerative neural condition in sheep. Prusiner named this infectious protein a *prion*. Prions also cause other neurologic diseases such as *kuru* and *Gerstmann-Strausler-Sheinker syndrome*.

However, scientists are still studying prions to learn their origins and how prions replicate and cause disease.

QUIZ

Match the virus—DNA or RNA:

1. Varicella-zoster virus _____

2. Cytomegalovirus _____

3. HIV _____

4. Eastern equine encephalitis _____

5. Herpes simplex virus type 1 _____

Match the scientist:

    A. Stanley Prusiner
    B. Wendell Stanley
    C. Martinus Beijerinck
    D. Carlton Gajdusek
    E. O. T. Diener

6. This person studied the tobacco mosaic disease. _____

7. This person showed that viruses could be regarded as chemical matter. _____

8. This person discovered viroids. _____

9. This person won the Nobel prize for his work with kuru. _____

10. This person proposed that proteins cause the neurologic disease scrapie. _____

chapter **13**

Epidemiology and Disease

In this chapter, you'll learn about how disease spreads when you read about epidemiology and how to control and prevent outbreaks by following a few basic rules.

CHAPTER OBJECTIVES

In this chapter, you will

- Begin to understand the goals of epidemiology
- Learn about the classification of disease
- Learn about infection
- Become familiar with disease transmission

Growing up, you were taught to share with friends and relatives. However, you probably shared a little too much when you shared your cold with your classmates. Pathogens such as bacteria and viruses are spread to others by touch or in droplets that float in the air.

You probably covered your nose with your hand and turned away when you felt a sneeze coming on. Most of the air droplets produced by the sneeze ended up on your hand, and the rest remained airborne, hopefully not moving in anyone's direction. However, microorganisms that caused you to sneeze were still on your hand and were transferred to anything you touched. This is the most common way germs are transferred among family members, friends, and strangers.

In this chapter you will learn about how disease is spread as well as how to control and prevent outbreaks by following a few basic rules.

What Is Epidemiology?

Epidemiology is the study of the distribution and cause of diseases or conditions of a population. The word originated from the two Greek words *epi* (meaning "among") and *demos* (meaning "people"), or "among the people." An *epidemiologist* is a scientist trained to identify and prevent diseases in a given population. Epidemiologists are concerned with the *etiology*, or specific cause, of a disease in a given population. Epidemiologists use this information to design ways to prevent and control outbreaks of disease.

Epidemiology is considered a branch of microbiology because microorganisms cause many diseases. It also can be considered a branch of ecology because of the relationship among pathogens, their hosts, and the environment. The science of epidemiology provides the methods and information that are used to understand and control outbreaks of diseases in human populations, making it important for public health. An epidemiologist is a person who is trained to identify and prevent diseases in a given population or a medical doctor who is trained to identify and treat diseases in individual people.

Epidemiologists are concerned with the *frequency* or *prevalence* of diseases in a given population. An epidemiologist will identify the factors that cause disease or how that disease is transmitted and how the spread of communicable and noncommunicable diseases can be prevented. The *incidence rate* of a disease is the total number of new cases seen within a calendar year. The *prevalence* of a disease is the number of people infected at any given time. The *prevalence rate* is the total number of old and new cases of a disease. Frequencies also are expressed as proportions of the total population. The *morbidity rate* is the state of illness or the number of people in a given population that are ill. This is expressed as the number of cases per 100,000 people per year. The *mortality rate* is the number of people that are dead or die as a result of a particular illness. This is measured as the number of deaths from a specific cause per 100,000 people per year.

Classification of Disease

Epidemiologists measure the frequency of diseases within a given population with regard to the geographic size of the area and the amount of damage the disease inflicts on the population. Diseases can be classified as endemic, sporadic, epidemic, or pandemic.

An *endemic disease* is a disease that shows an average or normal number of cases in a certain population. The number of people contracting the disease and the severity of the disease are so low that the disease raises little concern and does not constitute a health problem. An example is varicella-zoster virus infection (the virus that causes chickenpox). Chickenpox is an endemic disease that usually affects children and is seasonal. However, an endemic disease can give rise to epidemics.

A *sporadic disease* occurs when small numbers of isolated cases are reported. Sporadic diseases do not threaten the population.

An *epidemic disease* arises when the level of disease in a certain population exceeds the endemic level. This disease will cause an increase in the mortality rate and the rate of morbidity. The level of destruction will be so large that the disease will be a significant public health concern.

A disease becomes *pandemic* when it is distributed throughout the world. For example, in 2009, the swine flu (the flu associated with the H1N1 virus) reached pandemic proportions. Some experts consider HIV virus infection to be pandemic.

In a *common-source epidemic*, large numbers of the population are suddenly infected from the same source. These epidemics usually are attributed to a contaminated supply of water or improperly prepared or handled food. An example is people who eat contaminated chicken salad at a college cafeteria. Everyone who eats the chicken salad on that particular day will become infected and feel ill. The epidemic will subside very fast, though, as the source of infection is eradicated.

A *propagated epidemic* occurs from person-to-person contact. The disease-causing agent moves from a person who is infected to a person who is not infected. In a propagated epidemic, the number of new cases rises and falls much more slowly than in common-source epidemics, making the pathogen much harder to isolate and thus eliminate. An example is a flu virus.

Pathognomonic is a word that refers to the specific characteristics of disease. *Immunity* is the specific resistance to disease. *Virulence* is the degree of pathogenicity or the capacity of an organism to produce disease.

Pathology is the study of disease. It is derived from the Greek *pathos* (meaning "suffering") and *logos* (which means "science"). Pathology is that branch of study concerned with the structural and functional changes that occur in an organism owing to a disease-causing agent or pathogen. A *pathologist* is a scientist or physician who studies the cause of diseases, or *etiology*, and the *pathogenesis*, the manner in which a disease develops.

Infection Sites

The sites where a microorganism can infect a host organism are called *reservoirs of infection*. In these sites, a microorganism can maintain its ability to cause infection. Reservoirs of infection include humans, some animals, certain nonliving media, and inanimate objects.

Human Reservoirs

Humans make good reservoirs because they can transmit organisms to other humans. Certain disease-causing agents have an *incubation period* during which they are contagious and can spread the disease even before a person exhibits signs or symptoms. These disease-causing agents even can be contagious during the recovery period.

When a person becomes *symptomatic* (feeling sick), the individual seeks medical attention and receives treatment. In many cases, however, diseases are spread from a person with *subclinical findings*—the symptoms are very mild and not recognizable. These individuals can spread the disease to healthy people. *Asymptomatic* patients are a problem because infected people can infect other individuals without knowing that they are infected. These people who are "carriers" of disease are called *disease carriers*.

Carriers of disease are classified as

- *Subclinical carriers.* These individuals never develop clinical symptoms of the disease.

- *Incubatory carriers.* These individuals transmit the disease before becoming symptomatic.
- *Convalescent carriers.* These individuals are recovering from the disease, but they still can infect other people.
- *Chronic carriers.* These individuals develop chronic infections and transmit the infection for long periods of time.

Animal Reservoirs

Many microorganisms can infect both humans and animals. Many of these disease-causing agents use animals as reservoirs of infection to infect humans. Apes and monkeys are good examples of animals that serve as reservoirs for human infection because they are physiologically similar to humans. When an animal infects humans, the humans also can serve as reservoirs for the infection.

A disease that is transmitted from domestic and wild animals to humans is called a *zoonosis*. One example of a zoonosis is *anthrax*, which is a bacterial disease that causes infection in dogs, cats, cattle, and other domestic animals. Humans become infected by direct contact with the animals, their wool, or hides; contaminated soil; inhalation of spores; and ingestion of meat or milk. Another zoonosis is *rabies*, a virus that infects dogs, cats, skunks, wolves, and bats. Humans become infected through infected saliva in bite wounds. Humans and domestic animals also can be reservoirs for wild animals.

Nonliving Reservoirs

Examples of *nonliving reservoirs* are water and soil. Soil is a good reservoir for the bacterium *Clostridium tetani*, which causes the disease *tetanus*. Contaminated water infected with human or animal fecal matter can contain many disease-causing agents. An example is the bacterium *Vibrio cholera*, which causes *Asiatic cholera*, a disease caused by fecally contaminated water where sanitation is poor. This organism invades the intestines and causes severe vomiting, diarrhea, abdominal pain, and dehydration.

Disease Transmission

A disease must have a *portal of exit* from the reservoir and a *portal of entry* into the host. This is how diseases are spread and new cases of infection occur. Examples of portals of exit are the respiratory tract, digestive tract, urinary tract, skin, and in utero transmission. Diseases can spread by three different modes of transmission: contact transmission, vehicle transmission, and vector transmission.

Contact Transmission

Contact transmission of a disease-causing agent can either be *direct* or *indirect*. Direct contact transmission occurs from skin-to-skin contact, such as shaking hands, kissing, sexual contact, or making contact with open sores. Examples of diseases caused by direct contact transmission include herpes, syphilis, and staphylococcal infections.

Indirect contact transmission, or *vehicle transmission*, occurs when infection is spread through any nonliving, inanimate object. These contaminated inanimate objects are called *fomites* and include bedding, towels, clothing, dishes, utensils, glasses and cups, diapers, tissues, and even bars of soap. Examples of diseases caused by indirect contact transmission include the common cold, hepatitis B infection, and tetanus.

Droplet transmission is a form of contact transmission that occurs through sneezing, coughing, and speaking in close contact with an infected individual. Examples of diseases transmitting in this manner are pneumonia, influenza, the common cold, and whooping cough.

Vehicle Transmission

In *vehicle transmissions*, pathogens can be spread through the air and in water, food products, and bodily fluids (e.g., blood and semen). Airborne microorganisms come mainly from animals, plants, water, and soil. These microorganisms can transmit disease through air. They can travel 1 meter or more through an air medium to spread infection.

Airborne pathogens have the greatest chance of infecting new individuals when these individuals are crowded together indoors or in a climate-controlled building where heating and air-conditioning units regulate temperature and very little fresh air enters the building. Airborne pathogens can fall to the floor and combine with dust particles. This dust then can be stirred up by walking, dry mopping, or changing bedding and clothing. Examples of diseases that are transmitted by airborne transmissions and dust particles are measles, chickenpox, histoplasmosis, and tuberculosis.

Waterborne microorganisms that cause disease do not grow in pure water. They can survive in water with small amounts of nutrients but thrive in polluted water, such as water contaminated with fertilizer and sewage (which is rich in nutrients). Waterborne pathogens are usually transmitted in contaminated water supplies by either untreated or inadequately treated sewage. Indirect fecal-oral transmission of pathogens occurs when the disease-causing microorganism living in the fecal matter of one organism infects another organism. Bacterial pathogens infect the digestive system, causing gastrointestinal signs and symptoms. Examples of waterborne diseases are shigellosis and cholera.

Food-borne pathogens are normally transmitted through improperly cooked or improperly refrigerated food or unsanitary conditions. Improper hygiene on the part of food handlers also plays a key role in food-borne transmission. Food-borne pathogens can produce gastrointestinal signs and symptoms. Examples of food-borne diseases are salmonellosis, typhoid fever, tapeworm, and listeriosis.

Vector Transmission

Vector spread is the transmission of an infectious agent by a living organism to humans. Vectors include ticks, flies, and mosquitoes. These organisms are called *arthropods*. Vectors can transmit disease in two ways. First, mechanical vectors can passively transmit disease with their bodies. An example is the common housefly. These animals commonly feed on fecal matter. They then fly to feed on human food, transmitting pathogens along the way. Keeping mechanical vectors away from food preparation and eating areas is the best means of prevention. Remember, the fly that is walking across your picnic lunch may have just walked across dog or cat feces. Examples of a few diseases transmitted by mechanical vectors are diarrhea caused by *Escherichia coli* bacteria, conjunctivitis, and salmonellosis.

The second type of vectors are *biological vectors*, which can actively transmit disease-causing pathogens that complete part of their life cycle within the vector. In most vector-transmitted diseases, a biological vector is the host for a phase of the life cycle of the pathogen. An example of a host organism is a mosquito that infects a human with malaria. Other diseases caused by biological transmission vectors are yellow fever, plague, typhus, and Rocky Mountain spotted fever.

To cause infection, a microorganism must enter the body and have access to body tissues. The sites where microorganisms enter the body are called *portals of entry*. The portal of entry is similar to the portal of exit for the host to be susceptible for a certain disease. The portals of entry include the skin, digestive tract, respiratory tract, reproductive tract, and urinary tract. Microorganisms can invade tissues directly or cross the placenta to infect the fetus. Skin that is intact prevents most microorganisms from entering the body, although some enter the ducts of sudoriferous glands (sweat glands) and hair follicles to gain entrance into the body.

Fungal spores carried by animals, such as birds or insects, can invade cells on the surface of the body, and some even can invade other tissues. The larvae of some parasitic worms can work their way through the skin and enter tissues. An example of a parasitic worm is the hookworm.

Mucous membranes make direct contact with the external environment. This allows microorganisms to enter the body. Examples of mucous membranes are the eyes, nose, mouth, urethra, vagina, and anus. The respiratory tract is an area of the body where microorganisms typically enter on dust particles that are inhaled with air or in aerosol droplets. Microorganisms that infect the digestive tract are normally ingested with contaminated water or food or even from biting the nails of contaminated fingers.

Many genitourinary infections are the result of sexual contact. Skin that is not intact owing to injury, surgery, injections, burns, and bites makes it easy for invading microorganisms to penetrate body tissues. Common portals of entry are insect bites. Many parasitic diseases are caused by the bites of insects. Some diseases can affect the fetus through the placenta of an infected mother. Viruses such as the HIV virus, rubella (German measles), and the bacteria that cause syphilis behave in this way.

The transmission of disease by carriers causes epidemiologic problems because carriers usually do not know that they are infected and spread the disease, causing sudden outbreaks. Carriers can transmit disease by direct and indirect contact or through vehicles, such as water, air, and food.

Still Struggling

Having trouble remembering how diseases get transmitted. Here is a quick study:

Direct Transmission: Vector transmission—infection occurs through the transmission of infectious agents by a living organism to humans. *Contact transmission* (infections spread by skin-to-skin contact); *droplet transmission* (infectious through sneezing, coughing close, speaking); *biological vectors; across the placenta.*

Indirect Transmission: Vehicle transmission—infection is spread via air, water, or food. *Indirect contact transmission,* infections occur from nonliving inanimate objects such as fomites, food, water, or air.

The Development of Disease

In order for a pathogen to infect a host, there must be a susceptible host for the disease to be transmitted to. If a host's *resistance* is low (resistance is the ability to ward off disease), its susceptibility increases (its chances of becoming infected increase). Primary defense mechanisms of the body for resistance include intact skin (no cuts or abrasions), mucous membranes, a good cough reflex, normal gastric juices, and normal bacterial flora.

If a microorganism penetrates these defenses, the development of a disease process begins. First, there may be an *incubation period*. This is the time between the initial exposure and the start of the infection to the first appearance of signs and the feeling of symptoms. Different microorganisms have different incubation periods. An example is the Epstein-Barr virus, which causes infectious mononucleosis, which has an incubation period of two to six weeks.

The varicella-zoster virus, which causes varicella (chickenpox), has an incubation period of two weeks. The human immunodeficiency virus (HIV), the virus that causes AIDS, has an incubation period of 7 to 11 years. During this phase, the disease can be spread from the infected individual to noninfected individuals.

The *prodromal period* follows the incubation period. This period presents with mild symptoms.

The *period of illness* is the acute phase of the disease. Here, the individual presents with signs and symptoms of the disease. *Signs* are objective findings that an observer or physician can see. These are physical changes that can be measured. Examples of signs are fever, skin color or lesions, elevated blood pressure, inflammation, and paralysis.

Symptoms are subjective and cannot be seen by an observer. Symptoms present as changes in bodily functions, such as pain, numbness, chills, general fatigue, or gastrointestinal discomfort. It is in this period of the disease that white blood cells may increase, and the individual's immune system responds to combat the disease-causing pathogen. If the individual's defense mechanism of the immune system does not overcome the disease successfully, or if the disease is not treated properly, the person can die.

During the *period of decline*, the individual's signs and symptoms subside, and the person feels better. This period may take 24 hours to several days or even weeks. During this time, the individual is prone to secondary infections.

The *period of convalescence* is the phase where recovery has occurred. The body regains strength and is returned to a state of normality. During this phase, infection also can be spread.

Epidemiologic Studies

Epidemiologic studies began in 1855 with the work of English physician John Snow. Snow conducted studies relating to the cholera outbreak in London, England. Snow, through careful analysis of deaths related to cholera, case histories of victims, and interviews with survivors, traced the source of the epidemic to a water pump. He concluded that individuals who contracted cholera drank water contaminated with human feces. Snow's study and his method of analyzing where and when a disease could occur and the transmission of that disease within a given population gave way to a new approach in medical research and epidemiologic studies.

It was not until 1883 that the cholera bacterium *Vibrio cholerae* was identified by Robert Koch. After the studies of Snow, other investigators conducted epidemiologic studies. Epidemiologists now use three types of studies when determining the occurrence of disease: descriptive, analytical, and experimental.

Descriptive epidemiology is the collection of all data described in the occurrence of disease. These data include the number of cases, which portion of the population is infected, where the cases occurred, and information about the affected individuals (i.e., race, sex, age, occupation, marital status, and socioeconomic status).

Analytical epidemiology analyzes the cause of a disease and the effect of the disease in a given population. Here, epidemiologists compare a group of people who have the disease with people who do not have the disease.

Experimental epidemiology involves studies designed to test hypotheses for the source of a particular disease. Experiments are conducted on human or animal subjects to test the hypothesis. An example is the testing of an experimental drug designed to control a specific disease. A group of infected individuals is divided randomly so that one group receives the drug and the other receives a placebo. A *placebo* is a substance that has no effect on the individual receiving it, but the individual believes that he or she is receiving treatment. If the people who were treated with the drug recover faster than those taking the placebo, investigators can conclude that the drug treatment is effective.

Control of Communicable Diseases

There are different ways of controlling or limiting the spread of communicable diseases. These methods include isolation, quarantine, immunization, and vector control.

Isolation requires that a patient infected with a communicable disease be prevented from making contact with the general public. The Centers for Disease Control and Prevention (CDC) has designated five categories of isolation: strict, protective, respiratory, enteric, and wound and skin.

Quarantine requires the separation of animals and humans that have been infected or exposed to a communicable disease from the general public.

Immunization is an effective means of controlling the spread of communicable diseases by the use of safe vaccines. A *vaccine* is a preparation of killed, attenuated, inactivated, or fully virulent organisms that is administered to induce or produce artificially acquired active immunity.

Vector control is a good way of controlling the spread of infectious disease when the vector, such as rodents or insects, is identified. This vector's habitats and breeding grounds can be treated with insecticides and poisons. Also, barriers such as window screens, netting, and repellents can provide protection against bites and infection.

Nosocomial Infections

A *nosocomial infection* is an infection that is the result of a pathogen that was acquired in a hospital or clinical care facility. *Nosocomial* is derived from the Latin word *nosocomium*, which means "hospital." These are the diseases that a patient can obtain when he or she is being cared for in a hospital. These diseases also can affect the caregivers, such as the hospital staff, nurses, doctors, aides, and even visitors or anyone else who has contact with a hospital or medical facility.

The CDC estimates that among the patients who are admitted to hospitals, 5 to 15 percent acquire some type of nosocomial infection. Nosocomial infections result directly in over 20,000 deaths and indirectly in about 60,000 deaths annually.

Nosocomial infections result from either endogenous or exogenous sources. *Endogenous* infections are caused by pathogens that were brought into the hospital by the patient; the opportunistic pathogen is among the patient's own microbiota.

Exogenous infections are caused by organisms that enter the patient's body from the external environment. These organisms can be acquired from animate sources, such as hospital staff, other patients, or people visiting the hospital. Organisms also can come from inanimate sources, such as hospital equipment, intravenous and respiratory therapy equipment, catheters, computer keyboards, bathroom fixtures, doorknobs, soaps, and even certain disinfectants.

Who Is Susceptible?

A person who is susceptible to a nosocomial infection is one who has a compromised immune system. Compared with the general public, patients in hospitals have a lowered resistance to disease, thus making them more susceptible to infections. Many of these patients are said to be "compromised hosts"—they have breaks in their skin owing to accidental or surgical wounds (i.e., lesions, bedsores, or burns). Others may have compromised mucous membranes that line the respiratory tract, the digestive tract, or the urinary and reproductive system, making them more susceptible to disease-causing pathogens.

Prevention and Control of Nosocomial Infections

Most hospitals and clinical facilities have control measures and procedures aimed at preventing nosocomial infections. For a hospital to be accredited, it must have a designated person responsible for developing and implementing policies and procedures that would control infections and communicable diseases. This person can be a registered nurse or an epidemiologist.

Public Health Organizations: The Centers for Disease Control and Prevention

Public health agencies have been created in cities, countries, states, and at the federal level. The United States has recognized the importance of identifying and controlling infectious diseases since the eighteenth century. The Centers for Disease Control and Prevention (CDC) is a branch of the U.S. Public Health Service (USPHS). The CDC is located in Atlanta, Georgia, and is the central source of epidemiologic study and information. The CDC is responsible for the control and prevention of disease, public education, and occupational health and safety.

QUIZ

Name the carriers:

1. These individuals never develop clinical symptoms. _____

2. These individuals develop chronic infections and transmit the infection for long periods of time. _____

3. These individuals are recovering from the disease, although they still can infect other individuals. _____

4. These individuals transmit the disease before becoming symptomatic. _____

Match the transmission/infection:

   A. Contact transmission
   B. Droplet transmission
   C. Indirect contact transmission
   D. Vector transmission
   E. Vehicle transmission
   F. F. Nosocomial

5. Infectious pathogen that was acquired in a clinical care facility or hospital. _____

6. Transmission occurs through sneezing and coughing. _____

7. Transmission can occur through contact with infected towels and bedding. _____

8. Transmission occurs through spread of bodily fluids. _____

9. Transmission occurs via insects. _____

10. Transmission can occur from skin-to-skin contact. _____

14

Immunity

In this chapter, you'll learn about the immune system and the different immune system functions that protect your body from the ongoing attack of microbes.

CHAPTER OBJECTIVES

In this chapter, you will

- Learn about the immune system
- Be introduced to acquired immunity
- Learn about antigens
- Learn about antibodies

Your body is attacked by microorganisms 24 hours, 7 days a week. They find and burrow into every small crack and opening in your skin, looking to find a new home and nutrients so that they can sustain life and reproduce. Fortunately, you have your own Seal Team 6 that is called into action at the first sign of the invasion. This is your *immune system*.

You don't see these soldiers. In fact, you don't know that there is a battle going on until you experience the side effects of the fighting—runny nose, fever, and cough. These are some of the ways that your immune system counterattacks invading microbes.

The soldiers are called *B cells*, *T cells*, and *natural killer cells*. Their job is to recognize the attack and rip apart the invaders before any serious damage is done to your body. First, the microbe is surrounded; then it is destroyed. Sometimes, your Seal Team 6 needs help in the way of medication. Medication destroys enough of the attackers to even the playing field, enabling your immune system to finish the job.

What Is Immunity?

Immunity is the ability of an organism to make certain chemicals called antibodies and produce certain cells called lymphocytes to protect itself from disease-causing microorganisms. This ability occurs during the life of the organisms. *Resistance* is an organism's ability to defend itself against disease processes or pathogens.

There are two forms of resistance: *Nonspecific resistance* and *specific resistance*. Nonspecific resistance deals with a wide variety of responses against pathogens and plays a role in the first line of defense. The first line of defense prevents pathogens from entering the body. Specific resistance deals with the second line of defense or immunity, should the pathogen enter the body. *Immunity* is a specific defensive response of a host when a foreign substance or organism invades it.

An *immune system* is an organism's protection from invading organisms and foreign substances such as bacteria, viruses, fungi, helminths, protozoa, pollen, transplanted tissues, and insect venom. Parts of these substances are called *antigens* or *immunogens*. Antigens are seen as foreign to the body and can provoke a specific immune response in an organism.

The body reacts to the foreign substance (*antigen*) by forming antibodies. *Antibodies* are proteins that are made by the body in response to an antigen and can combine specifically with only that antigen.

The immune system recognizes a body or substance within the organism as either self or nonself. *Self* is any component that belongs to the organism. *Nonself* is anything that doesn't belong to the organism. An antigen is recognized as nonself by the immune system.

An antigen causes the organism to form *antibodies* and *specialized lymphocytes* that target the specific antigen. If the antigen invades again, these antibodies and specialized lymphocytes attack the antigen, destroying the antigen or making it inactive. This is called an *immune response* against the antigen.

Cells that become cancerous are recognized as foreign and can be destroyed. Cancerous cells, once established as a tumor, may have a dreadful or fatal effect on the organism because the immune system is not able to fight them in such large numbers.

Acquired Immunity

The development of antibodies and specialized lymphocytes is called *acquired immunity* (see Fig. 14-1) because it is acquired over an organism's life through

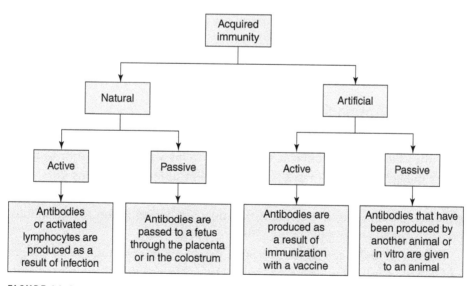

FIGURE 14-1 • Types of Acquired Immunity. This illustration can be used as a guide to the text description of acquired immunity. The details of the immune system are astonishingly complex; many events have been omitted so that the overall organization of the immune response can be clearly seen. (From Prescott et al., *Microbiology,* 6th ed., McGraw-Hill, 1996.)

natural and artificial means. It is the protective defense mechanism an organism develops against foreign substances and microorganisms.

There are two types of acquired immunity. These are active and passive. When an individual is exposed to a disease-causing microorganism or foreign substance, the person's own immune system responds by making its own antibodies and lymphocytes. This is *active immunity*.

In *passive immunity*, already-made antibodies are introduced or passed on to the individual. The individual does not make his or her own antibodies.

Naturally acquired active immunity occurs when an individual is exposed to an infectious disease. The individual's immune system responds by making its own antibodies and lymphocytes (T cells and B cells) on its own.

Naturally acquired passive immunity occurs when antibodies (IgG) are made by a mother and passed on to her fetus through the placenta. IgA antibodies are also passed to the baby in the first secretion of breast milk, called *colostrum*, during breast-feeding.

Artificially acquired active immunity occurs when an individual is given a vaccine. A *vaccine* is a substance that contains the weakened or dead organism, which stimulates the immune response but does not cause major illness. The body remembers the antigen with memory cells the next time it is exposed to the antigen.

Artificially acquired passive immunity occurs when antibodies are developed outside the individual and injected intravenously into the body. This form of immunity helps the body's own defenses in combating infection.

Serum and Antibodies

Serum is the liquid portion of blood. *Electrophoresis* is a laboratory technique that introduces electric current to cause proteins within the serum to move across the gel at different rates, which represent different globulins.

There are four globulins. These are called *alpha 1, alpha 2, beta*, and *gamma globulins*. *Gamma globulin* contains the most antibodies. A serum rich in antibodies is called either *gamma globulin* or *serum globulin*.

Gamma globulin can be taken from a person who is immune to an antigen and injected into a person who lacks the antigen, who then immediately becomes immune to the antigen. However, the immunity lasts for about three weeks, at which time the antibodies degrade.

Still Struggling

Having trouble with the types of immunity? Let's review them again. In *naturally acquired active immunity*, a person is infected with an organism, makes antibodies, and naturally fights the infection. In *naturally acquired passive immunity*, a person receives antibodies from his or her mother. In *artificially acquired active immunity*, a person receives a vaccine with the weakened or dead infectious organism, and his or her body makes antibodies and becomes ready if the person becomes exposed to the actual infectious organism. In *artificially acquired passive immunity*, a person receives antibodies directly to fight the infection.

Types of Immunity

Antibody-Mediated Immunity

Antibody-mediated immunity, which is also known as *humoral immunity*, uses antibodies in extracellular fluids, such as mucus secretions, blood plasma, and lymph, to combat antigens. These antibodies, produced from *B cells*, which are also known as *B lymphocytes*, primarily attack bacteria, bacterial toxins, and viruses that invade body fluids. They also attack transplanted tissues.

Antibody-mediated immunity was discovered by German scientist Emil von Behring at the turn of the twentieth century in his quest to create an immunization against diphtheria. Behring called this *humoral immunity* because the medical community called body fluids *humors* (Fig. 14-2).

Cell-Mediated Immunity

Cell-mediated immunity involves specialized lymphocytes called *T cells*, also known as *T lymphocytes*, that attack foreign organisms as well as abnormal body cells, rather than using antibodies. T cells are also effective against helminths, fungi, and protozoa. In addition, T cells regulate aspects of the immune system.

Cell-mediated immunity was explained by Russian biologist Elie Metchnikoff, who in the early 1900s noticed that phagocytic cells were much more effective in animals that were immunized. This immunity was used in the mid-twentieth century to protect people against tuberculosis (see Fig. 14-2).

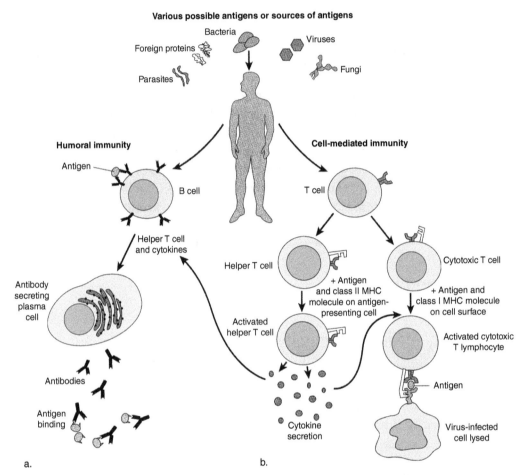

FIGURE 14-2 • The Humoral and Cell-Mediated Branches of Immunity. (a) In the humoral branch, B cells interact with antigens and differentiate into antibody-secreting plasma cells. The antibody binds to the antigen and tags it for destruction by other mechanisms. (b) In the cell-mediated response, subpopulations of T cells are activated by antigen presented in the presence of a major histocompatibility complex (MHC). Activated helper T cells respond by producing a group of specific chemicals called cytokines that facilitate both the humoral and cell-mediated responses. Cytotoxic T cells respond to the antigen by developing into cytotoxic T lymphocytes that kill virus-infected cells and other altered cells. B cells and T cells also differentiate into memory cells (not shown) that are involved in subsequent responses. B cells and T cells also differentiate into memory cells involved in subsequent secondary responses (not shown). (From Prescott et al., *Microbiology*, 6th ed., McGraw-Hill, 1996.)

A Closer Look at Antigens

An antibody identifies its corresponding antigen by one or more regions on the antigen known as the *epitopes*, which are also called *antigenic determinants*. The epitope must be the right size, shape, and chemical structure for the antibody to bind to the epitope and then proceed to disable or destroy the antigen.

Antigens tend to have a molecular weight of 10,000 or more, yet some foreign substances may have a lower molecular weight and are not antigens. They are called *haptens* and must attach themselves to a large carrier molecule to become antigenic. Antibodies only attack the hapten and not the carrier molecule.

The antibiotic Penicillin behaves like a hapten in some people. Penicillin does not have an antigenic effect in most people. However, when penicillin attaches to serum proteins, an allergic reaction results in some people. These people are said to be allergic to penicillin. An allergic reaction is a typical immune response.

Antigens can be proteins, large polysaccharides, lipids, or nucleic acids. However, antigens that are lipids and nucleic acids must be combined with proteins and polysaccharides; otherwise, they are not antigens.

A Closer Look at Binding

Antibodies are a class of proteins known as *immunoglobulins* (Igs). An antigen can cause the production of different antibodies if the antigen has several recognized epitopes. Epitopes or antigenic determinants are known as *antigen-binding sites*. The number of antigen-binding sites is called the antibody's *valence*. There are at least two antigen-binding sites on each antigen where a human antibody can bind. This is referred to as a *bivalent antibody* because the antibody's valence value is two.

The structure of a bivalent antibody consists of four protein chains that are named after their relative molecular weights. These are two *light (L) chains* and two *heavy (H) chains*.

These protein chains are joined together to form a Y-shaped molecule that is flexible enough to form a T-shaped molecule. There are two regions of the protein chain, the *variable (V) region* and the *constant (C) region*.

The variable region is located at the ends of the arms of the Y. These are the sites where the antibody binds with an antigen. The variable region is a three-dimensional structure of amino acid sequences whose structure reflects the epitopes of a specific antigen.

The constant region is the stem of the Y and is called the *Fc region*. The Fc region binds adjacent complementary antibodies if both antigen-binding sites are attached to an antigen; otherwise, the adjacent complementary antibodies are free to attach and react with an antigen.

There are five types of constant regions, each associated with the five classes of immunoglobulin. The class involved in a response depends on the type of invading antigen, the portal of entry involved, and the antibody function required.

IgG

IgG immunoglobulin neutralizes bacterial toxins and attacks circulating bacteria and viruses by enhancing the effectiveness of phagocytic cells. Nearly 80 percent of the antibodies in serum are IgG. IgG can cross blood vessel walls and the placenta and can enter tissue fluids.

IgM

IgM immunoglobulin is the first antibody to respond to an antigen or initial infection and makes up about 10 percent of antibodies in serum. IgM is relatively large in size and has a pentamer structure of five monomers bonded together by a *joining (J) chain*. This chain is a polypeptide. Because of its size, IgM remains in blood vessels and not in tissue fluids. IgM responds to the ABO blood group antigens and enhances the effectiveness of phagocytic cells. When initial exposure to an antigen occurs, IgM antibodies are the first to appear.

IgA

IgA immunoglobulin is the most common antibody in body secretions and in mucous membranes. It makes up about 15 percent of serum antibodies. IgA protects infants from gastrointestinal infections and fights antigens that affect the respiratory tract. Plasma cells in the mucous membrane form secretory IgA, which then is passed through the mucosal cell and attacks antigens such as bacteria and viruses on the mucosal surface. IgA is short-lived.

IgD

IgD immunoglobulin is found in blood and lymph fluid and is an antigen receptor on the surfaces of B cells. Scientists do not know the function or importance of IgD.

IgE

IgE immunoglobulin binds to basophil cells and mast cells that release chemical mediators such as histamine that cause an allergic reaction. When pollen is the antigen, the allergic reaction is referred to as *hay fever*. IgE is less than 1 percent of serum antibodies. IgE is also known to provide protection against parasitic worms.

B Cells (B Lymphocytes)

B cells are cells that develop from stem cells in the bone marrow and liver of fetuses. They are transported to the lymph nodes and spleen, where they use *antigen receptors*, also known as *antigen-binding sites*, on the cell's surface to seek out antigens.

Once an antigen is detected, the B cell, along with T cells, activates a special group of lymphocytes that produce antibodies used in the antibody-mediated immune response. T cells do not make antibodies. When B cells come in contact with extracellular antigens, the B cells transform into *plasma cells* that produce antibodies at about 2,000 antibodies per second to combat that antigen.

Memory cells are also produced when a B cell is stimulated by an antigen. A memory cell provides the organism with long-term immunity to the antigen.

B cells react to one kind of antigen that is referred to as its *complementary antigen*; they are able to identify that antigen because antigen receptors bind to one specific antigen. Here's how it works: Once the antigen binds to the antigen receptor, the B cell replicates by clonal selection, producing numerous "clone" cells and specific antibodies. These antibodies attach to the antigen at an antigen-binding site to form an *antigen-antibody complex*. This complex is very specific. However, when there are large quantities of antigens, the antigens attach to antibodies where they do not exactly fit. This makes for less than perfect matches for the antigen-antibody complex. These antibodies are said to have less of an attraction for the antigen.

B cells undergo apoptosis if the B cell does not come in contact with an antigen. *Apoptosis* is a programmed death of the B cell that causes phagocytes to remove the cell from the organism.

Strategies for Combating Antigens

The formation of antigen-antibody complexes is useful in the response to infectious organisms or foreign substances because they remove the infectious agent from the body. Removal of antigens by bound antibodies occurs by several mechanisms: agglutination, opsonization, neutralization, antibody-dependent cell-mediated cytotoxicity, and the activation of complement.

- *Agglutination* clumps together antigens, making them nonfunctional and increasing their chance of being ingested (phagocytosis).

- *Opsonization* coats the antigen with antibodies to make it easy for phagocytic cells to ingest and lyse them.

- *Neutralization* blocks antigens from attaching to targeted cells, thereby neutralizing the antigen.
- *Antibody-dependent cell-mediated cytotoxicity* coats the foreign cell with antibodies. Nonspecific immune cells then destroy the foreign cell from the outside. This is used for organisms that are too large for phagocytic cells to ingest.
- The *activation of complement* is used when infectious agents are coated with reactive proteins that cause IgG and IgM antibodies to attach to the agent, resulting in *lysis* of the cell membrane and ingestion by phagocytes.

Lasting Immunity

When antigens are first encountered, the *primary immune response* occurs, causing an increase in the antibody titer. The *antibody titer* is the amount of antibodies in the serum of the infected organism. There are no or undetectable levels of antibodies when the antigen first attacks the organism. The antibody titer increases gradually and then declines as the antigen is destroyed or neutralized.

When antigens are encountered for the second time, the *secondary immune response* occurs, causing memory cells to quickly transform into plasma cells that produce antibodies. The secondary immune response is also known as the *anamnestic response* or *memory response*. A *memory cell* is a B lymphocyte that was generated in the primary immune response but did not become an antibody-producing plasma cell at that time.

Antibodies Used for Diagnosing Diseases

Antibodies are useful in diagnosing diseases because a particular antibody is produced only if a complementary antigen is present in the organism. This is useful in identifying an unknown disease-causing pathogen.

Antibodies can be produced in the laboratory by a clone of cultured cells that make one type of antibody. These antibodies are called *monoclonal antibodies*. Malignant cells of the immune system called *myeloma cells* are used because they divide forever. This is why malignant cells (cancer cells) are so devastating to our bodies. These myeloma cells then are mixed with lymphocytes that have been designed to produce a specific antibody. When these cells are mixed together, they fuse to become one cell called a *hybridoma*. Hybridomas divide indefinitely because they have the gene from the myeloma cell. They can produce large amounts of antibodies because they have the gene from the lymphocyte.

Monoclonal antibodies are used to diagnose streptococcal bacteria and chlamydial infections. Some over-the-counter pregnancy tests use monoclonal antibodies to detect the hormones found in urine during pregnancy.

Chemical Messengers

Cells in the immune system communicate with each other by using chemical messengers that send signals to trigger activities. These chemical messengers are known as *cytokines*. There are approximately 200 known cytokines.

Cytokines used for communication between leukocytic cells are called *interleukins*. Table 14-1 lists important interleukins.

Cytokines are used as therapeutic agents to combat disease. For example, interleukin-1 is used to stop blood flow to tumors in animals, thereby killing the tumor.

T Cells (T Lymphocytes)

T cells develop from stem cells in bone marrow and migrate to the thymus gland, where they mature. They then migrate to the lymphatic system and circulatory system to begin their fight against antigens. The T stands for *thymus gland*. Once the organism reaches late adulthood, the ability to create new T cells diminishes, resulting in a weaker immune system as the organism ages.

A T cell attacks specific antigens that are on the surface of the cell. T cells can recognize these antigens after the cell containing these antigens is processed by *antigen-presenting cells* (APCs). After the antigen is processed by the APC, fragments of the antigen are placed on the surface of the antigen-presenting cell

TABLE 14-1 Important Interleukins

Interleukin	Description
Interleukin-1	Stimulates T cells
	Attracts phagocytes in an inflammatory response
Interleukin-2	Stimulates B-cell production
	Stimulates T-cell production
Interleukin-8	Attracts phagocytes to the inflammation site
Interleukin-12	Stimulates the differentiation of CD4-type T cells

with the *major histocompatibility complex* (MHC) molecule. This molecule is made up of a group of proteins that are unique to a person and are used to distinguish self from nonself. Cells that are antigen-presenting cells (APCs) would include macrophages and dendritic cells.

When an antigen receptor encounters fragments of the complementary antigen, the T cell transforms into an *effector T cell* that carries out the immune response. An effector T cell is an *antigen-stimulated cell*. Some T cells attack the antigen in a primary immune response, whereas others become memory cells and take on a secondary immune response role when the antigen is encountered later on.

There are three types of T cells, each identified by characteristics of their surface molecules. These are

- *Helper T (T_H) cells or CD4 cells*. These cause the formation of cytotoxic T cells, activate macrophages, produce cytokines, and are essential to the formation of antibodies by B cells.
- *Cytotoxic T (T_C) cells or CD8 cells*. These destroy cells that have been infected by viruses and bacteria.
- *Regulatory T cells (Treg)*. These turn off the immune response when there are no antigens.

T cells are also identified by their surface receptors, called *clusters of differentiation* (CDs). There are two types of CD cells. These are

- *CD4 cells*. Helper T cells.
- *CD8 cells*. Cytotoxic T cells and suppressor T cells.

Macrophages and Natural Killer Cells

Macrophages are phagocytic cells that ingest antigens. They are in a resting state until they receive a cytokine from the helper T cell, at which point they become large and ruffled and ready to attack the antigens. Macrophages destroy virus-infected cells and bacteria located within cells. They also eliminate some cancerous cells.

Natural killer (NK) cells are lymphocytes that destroy other cells such as tumor cells. Natural killer cells are always active and searching for an infected cell. These are different from other cells in the immune system, which become activated only when stimulated by an antigen.

QUIZ

Name the interleukin:

1. These interleukins attract phagocytes to the inflammation site. _____

2. These interleukins stimulate T cells and attract phagocytes in an inflammatory response. _____

3. These interleukins stimulate the production of both T and B cells. _____

4. These interleukins stimulate the differentiation of CD4-type T cells. _____

Match the T cell:

    A. Helper T cell
    B. Cytotoxic T cell
    C. Delayed hypersensitivity T cell
    D. Suppressor T cell
    E. CD4 cell
    F. CD8 cell

5. These T cells turn off the immune response when there are no antigens. _____

6. These T cells are essential to the formation of antibodies by B cells. _____

7. These T cells are associated with allergic reactions. _____

8. These T cells destroy cells that have been infected by a virus. _____

9. These clusters of differentiation are a type of helper T cell. _____

10. These clusters of differentiation are a type cytotoxic T cell. _____

15

Vaccines and Diagnosing Diseases

In this chapter, you'll learn about how the immune system works and about different ways vaccines can protect you from invading microorganisms.

CHAPTER OBJECTIVES

In this chapter, you will

- Learn what a vaccine is
- Become familiar with types of vaccines
- Learn how to develop a vaccine
- Examine how diseases are diagnosed

Who would volunteer to receive an injection if you felt well? Many of us do every year if the injection is a vaccine. A vaccine reduces the likelihood that you will come down with the flu or other disease caused by a microbe because it strengthens your immune system.

Think of a vaccine as sharing intelligence about the enemy with your Seal Team 6 (your immune system). Seal Team 6 can fight most microbes, but learning to identify the microbe sooner helps Seal Team 6 to use special tactics to quickly destroy the enemy. The vaccine tells your Seal Team 6 about a particular microbe.

The vaccine contains a very small amount of the microbe—not enough to cause disease but enough so that the immune system can create antibodies that fight the microbe. Think of an antibody as specially trained Seal Team 6 designed to battle the microbe.

Once antibodies are created, they patrol your bloodstream looking for that particular microbe. At the first sign, they attack and get rid of the microbe before it can cause you any trouble.

What Is a Vaccine?

A *vaccine* is a suspension that contains a part of a pathogen that induces the immune system to produce antibodies that combat the antigen. The concept of a vaccine stems from the variolation process that was used in eighteenth-century England to protect people from smallpox.

The *variolation process* involves placing a needle tip of the smallpox organism in the vein of a patient. Nearly all such patients contracted a mild case of smallpox, which left them with antibodies that protected them from contracting the disease in the future. Half the patients who contracted smallpox died. By contrast, only 1 percent of those who received the variolation process died.

Edward Jenner noticed that dairymaids who contracted cowpox, which is related chemically to smallpox, were immune to smallpox. Jenner discovered that injecting cowpox into the skin of a healthy person prevented that person from developing smallpox. Jenner's discovery enabled Louis Pasteur to develop the technique of creating vaccines.

The injection of an antigen induces the primary and secondary immune responses in the patient. The *primary immune response* produces antibodies, and the *secondary immune response* produces memory cells that attack future invasions of the antigen (see Chapter 14).

Vaccines play an important role in controlling the spread of viruses. A virus cannot be treated with antibiotics. However, you can minimize catching the flu by getting a flu shot, which is a vaccine against a particular strain of flu virus.

Vaccines also prevent bacterial infections such as typhoid but are not as effective on bacteria as they are on viruses. However, bacterial infections are usually treatable with antibiotics, which is a common method of combating bacterial diseases.

Scientists also use vaccines to provide herd immunity to a population. *Herd immunity* requires that most—not all—of the population be immunized to prevent an epidemic of a disease. An outbreak of a disease would be isolated to a small percentage of the population and therefore have a minimal effect.

Types of Vaccines

There are six types of vaccines. These are

- *Attenuated whole agent.* These vaccines are designed for people who have a normal immune system. The attenuated whole-agent vaccine uses weakened living microbes to mimic the real infection to produce 95 percent immunity over a long term without the need of a supplemental vaccination called a *booster*. Common attenuated whole-agent vaccines include those for tuberculosis bacillus, measles, rubella, Sabin polio, and mumps. There is a risk that live microorganisms can regain their strength, resulting in the patient contracting the disease.

- *Inactivated whole agent.* These vaccines also are designed for people who have a normal immune system. The inactivated whole-agent vaccine uses dead microbes that were killed by phenol or formalin. Common inactivated whole-agent vaccines include those for pneumonia, Salk polio, rabies, influenza, typhoid, and pertussis (commonly known as *whooping cough*).

- *Toxoids.* The toxoid vaccine is made of toxins produced by a virus or bacteria that has been inactive. They are then used against toxins that are produced by a disease-causing microorganism. Patients require a booster vaccination every 10 years because the toxoid vaccine does not provide lifelong immunity. Common toxoid vaccines include those for diphtheria and tetanus.

- *Subunit.* These vaccines have few side effects. The subunit vaccine uses fragments of a microorganism to create an immune response. Subunit vaccines produced by using genetic engineering techniques to insert the genes of an antigen into another organism are called *recombinant vaccines*. Common subunit vaccines include those for hepatitis B.

- *Conjugated*. These are fairly new in development and are designed for children under 24 months of age whose immune system normally does not respond well to vaccines. The vaccine is produced by combining polysaccharides found on a pathogen with a protein. This combination is then recognized by the immune system.

- *Nucleic acid*. These vaccines are in the animal testing stage. The nucleic acid vaccine, which is also called a *DNA vaccine*, contains plasmids of naked DNA and is designed to produce protein that stimulates an immune response. The nucleic acid vaccine has a strong effect on large parasites and viruses.

Still Struggling

Still having difficulty with vaccines? This summary can help:

Attenuated whole-agent vaccines are designed for people who have a normal immune system. Common attenuated whole-agent vaccines include those for tuberculosis, measles, rubella, Sabin polio, and mumps. *Inactivated whole-agent vaccines* also are designed for people who have a normal immune system. Common inactivated whole-agent vaccines include those for pneumonia, Salk polio, rabies, influenza, typhoid, and pertussis. *Toxoid vaccines* are made of toxins produced by a virus or bacteria that have been inactive. Patients will require a booster vaccination because these vaccines do not provide lifelong immunity. Common toxoid vaccines include those for diphtheria and tetanus. *Subunit vaccines* have few side effects and use fragments of a microorganism to create an immune response. Common subunit vaccines include those for hepatitis B. *Conjugated vaccines* are fairly new in development and are designed for children under 24 months of age. The vaccine is produced by combining a polysaccharide with a protein. *Nucleic acid vaccines* are in the animal testing stage. These vaccines are also called *DNA vaccines*. They have a strong effect on large parasites and viruses.

Developing a Vaccine

Vaccines are developed by cultivating a large quantity of a pathogen, which is a disease-causing organism. Some pathogens, such as the rabies virus, can be cultivated in animals. For example, a chick embryo is used commonly to grow viruses and is the method used to develop the influenza vaccine.

With the development of cell culture techniques, that use cells from humans and primates, large-scale viral growth is possible. Scientists use recombinant vaccines because they do not need an animal host to grow the microorganism. An example is the hepatitis B vaccine.

Diagnosing Diseases

The reaction between an antibody and its complementary antigen is used in conjunction with observing symptoms to diagnose a disease. You probably have seen this diagnostic technique used when you were tested for tuberculosis. This test required that a suspension of *Mycobacterium tuberculosis* be injected into your skin. If the site becomes a red raised welt within a couple of days (48 hours), then you have tested positive for tuberculosis. The redness is a reaction between antibodies and antigen.

Scientists use eight types of reactions (antigen-antibody immune testing) to diagnose diseases. Each determines whether a specific antibody or a specific antigen is present based on its complementary antibody or antigen. These include

- *Precipitation Tests*. IgG or IgM antibodies are combined with soluble antigens. If the antibody and antigen are in an optimal ratio, they form an antigen-antibody complex called a *lattice*. The reaction occurs immediately, but the lattice may not form until minutes after the reaction begins. There are three commonly used precipitation reaction tests. These are
  ○ *Precipitin ring test*. A ring appears in the area of the optimal ratio, which is called the *zone of equivalence*.
  ○ *Immunodiffusion test*. A line is visible in the area of the optimal ratio.
  ○ *Immunoelectrophoresis test*. This uses electrophoresis to identify separated proteins in human serum. The test is called the *Western blot test* and is used in AIDS testing.
- *Agglutination reaction*. Uses cells that possess antigen molecules or soluble antigens which are introduced to antibodies to form an aggregate reaction

called *agglutination* (clumping) of cells. There are two types of agglutination reaction tests. These are

- *Direct agglutination test*. This uses a plastic microtiter plate to detect antibody reaction with large cellular antigens. The microtiter plate contains a series of wells. Each well has the same amount of antigen and successively diluted serum containing antibody. The direct agglutination test measures the titer antibodies in the serum. The titer is lower at the onset of the disease and higher later in the disease.
- *Indirect agglutination test*. Soluble antigens are absorbed onto the surfaces of latex beads and then allowed with antibodies. This has a relatively short reaction time, enabling a diagnosis in 10 minutes. The indirect agglutination test is used commonly to diagnose streptococci, which cause sore throats.

- *Hemagglutination Tests*. This type of reaction uses red blood cell surface antigens and complementary antibodies to determine whether red blood cells clump in reaction to the antibodies. This test is used in typing blood and in diagnosing infectious mononucleosis. If the antigen is a virus, the reaction is called *viral hemagglutination*.

- *Neutralization Tests reaction*. This reaction uses antibodies as an antitoxin to block the extoxin or toxoid of bacteria or a virus. *Extoxin* is an active toxin, and *toxoid* is an inactive toxin. A serum containing the antibody for a particular antigen is placed in a cell culture that contains cells and the antigen. If the antigen does not destroy (*cytopathic effect*) cells, then the extoxin is neutralized. Therefore, the antigen is complementary to the antibody. Tests that use the neutralization reaction are called *in-vitro neutralization tests*. One such test is the *hemagglutination inhibition test*, which is used to diagnose a number of infections, including measles, influenza, and mumps.

- *Complement-fixation Tests*. This reaction uses a group of serum proteins called *complement* to bind with the antigen-antibody complex.

- *Fluorescent antibody (FA) Tests*. This reaction combines fluorescent dyes with an *antibody*, making the antibody fluoresce when exposed to *ultraviolet light*. It is used to detect a specific antibody in a serum. The fluorescent antibody reaction is used to test for rabies. There are two types of fluorescent antibody reaction tests. These are

- *Direct FA test*. The antigen and the fluorescein-labeled antibodies are combined on a slide and then incubated. The slide is washed to remove any antibodies that did not attach to the antigen and then observed under a fluorescence microscope for yellow-green fluorescence. This test is used to identify antigen in a tissue. It is also used to detect rabies, *Mycobacterium tuberculosis,* and yeast infections.

○ *Indirect FA test.* The antigen and the serum containing the antibody are combined on a slide. Fluorescein-labeled anti–human immune serum globulin (antiHISG) is also added to the slide, which reacts to human antibody on the slide. The slide is incubated, washed, and then observed using a fluorescence microscope. The presence of fluorescence indicates that the person has serum antibodies against the antigen.

- *Enzyme-linked immunosorbent assay (ELISA).* This is also a labeled antibody test. This test uses an enzyme that reacts with its substrate to produce a color change indicating a positive test. These are

○ *Direct ELISA.* The objective is to identify the antibody for a specific antigen. The antibody is placed in microtiter plate wells, where it adsorbs to the *surface of the well.* The antigen then is placed in each well. Wells then are washed. The antigen that reacts to the antibody will be retained and remain there after the washing. Next, a second antibody for the antigen is added and adsorbed to the well wall. The antigen is sandwiched between two antibody molecules if the test is positive; otherwise, the test is negative. Direct ELISA is used to detect drugs in urine.

○ *Indirect ELISA.* The objective is to identify the antigen for a specific antibody. A known antigen is placed in microtiter plate wells, where it adsorbs to the surface of the well. The antibody is placed in each well and will bind to the antigen to form an antigen-antibody complex. The wells are washed to remove antibodies that did not bind. AntiHISG that has been linked with an enzyme then is placed into the well and reacts to the antigen-antibody complex in the well. All unbound HISG is rinsed. A substrate for the enzyme is added. If there is a color change, then the test is positive.

- *Radioimmunoassay (RIA).* This uses radioactive tags to mark antigens and antibodies. Samples then are scanned to determine the presence of the tag.

QUIZ

Match the test:

    A. Immunoelectrophoresis test
    B. Direct agglutination test
    C. Indirect agglutination test
    D. Hemagglutination reaction
    E. Hemagglutination inhibition test
    F. Neutralization reaction
    G. Direct FA test
    H. Indirect FA test
    I. Direct ELISA test
    J. Radioimmunoassay test

1. This test measures the titer antibodies in serum. _____

2. In this test, the antigen being tested and fluorescein-labeled antibodies are combined and added to a slide. _____

3. In this test, the antigen being tested, the serum containing the antibody, and antiHISG are combined and added to a slide. _____

4. In this test, soluble antigens are adsorbed onto the surface of latex spheres, where they react to antibodies. _____

5. This test is used to detect drugs in urine. _____

6. This test uses radioactive tags to mark antigens and antibodies. _____

7. This test is used to diagnose measles, influenza, and mumps. _____

8. This reaction uses antibodies as an antitoxin to block the exotoxin or toxoid of bacteria or a virus. _____

9. This test is called the *Western blot test* and is used in AIDS testing. _____

10. This test is used in blood typing. _____

chapter 16

Antimicrobial Drugs

In this chapter, you'll learn about antimicrobial drugs and how antimicrobial drugs can destroy pathogenic microorganisms.

CHAPTER OBJECTIVES

In this chapter, you will

- Learn about chemotherapy
- Begin to understand the bacteriostatic strategy
- Learn about metabolites
- Appreciate antimicrobial drugs

At the first sign of feeling sick, you probably go to the doctor and ask for a pill to make you feel better. You are feeling the side effects of the battle between the microbe and your Seal Team 6—your *immune system*. In theory, your immune system should be able to make you feel better as long as it is not compromised. However, you are probably going to feel miserable for a while until the microbes are killed.

Alternatively, the doctor can prescribe antimicrobial medication that helps to destroy the microbe quicker than your immune system. *Medication* is a chemical agent called a *drug*, and taking the medication is called *undergoing chemotherapy*. The term *chemotherapy* is used commonly in relation to cancer treatment, where drugs are used to destroy your own rapidly growing cells. Chemotherapy to fight pathogenic microorganisms is used to destroy only the microorganisms themselves.

Chemotherapeutic Agents: The Silver Bullet

When we hear the term *chemotherapy*, we tend to think of an ongoing treatment for cancer. While this is true, chemotherapy is any treatment that introduces a chemical substance into the body to destroy a pathogenic microorganism. The chemical substance is a *chemotherapeutic agent*.

The chemotherapeutic agent must do two things. First, it must cause only minimal damage (or no damage) to host tissues. *Host* is the term scientists use to refer to the patient who is receiving chemotherapy; *host tissues* are tissues of the patient's body. This simply means that the chemotherapeutic agent must not injure the patient or, if there is injury, that the injury is minimal, and the patient's body will regenerate the destroyed tissues once chemotherapy is finished.

The second thing that the chemotherapeutic agent must do is destroy the pathogenic microorganism that is causing the disease. The way in which a chemotherapeutic agent destroys a pathogenic microorganism is called the chemotherapeutic agent's *action*. The pathogenic microorganism that is attacked by the chemotherapeutic agent is called the chemotherapeutic agent's *target*.

There is generally one of two actions that a chemotherapeutic agent takes when combating a target. Using bacteria as an example: one is to kill the bacteria outright, which is referred to as *bactericidal* action; the other is to inhibit the growth, which is called a *bacteriostatic* action. You'll learn more about bactericidal and bacteriostatic actions throughout this chapter.

A Look Back

The idea of chemotherapy was the brainchild of Paul Ehrlich, a German scientist, who in the early twentieth century predicted that chemotherapeutic agents could be used to treat diseases that were caused by microorganisms. Ehrlich based his prediction on the result of a wayward experiment. He tried to stain only the bacteria in a tissue sample without staining the tissue.

This became a major hurdle in the discovery of chemotherapeutic agents. Finding a chemotherapeutic agent to kill a pathogenic microorganism wasn't difficult, but chemotherapeutic agents also harmed and sometimes killed the patient, too. The challenge was to discover a chemotherapeutic agent that cured the disease and did not kill or severely injure the patient.

A breakthrough came in 1929 when Alexander Fleming was growing *Staphylococcus aureus*, a bacterium, in a Petri dish. A colony of mold contaminated the Petri dish and surrounded the *S. aureus* and consequently prevented the bacteria from growing. The mold was *Penicillium notatum*. Fleming was able to isolate the part of the *P. notatum* that stopped the growth of the *S. aureus*, which is referred to as the *active compound*. Fleming named this active compound *penicillin*.

The action of a compound to inhibit the growth of a microorganism is called *antibiosis*. From the word *antibiosis* comes the word *antibiotic*, which is any substance that a microorganism produces, in small amounts, that inhibits the growth of another microorganism. Therefore, penicillin is an antibiotic produced by the *P. notatum* microorganism that inhibits the growth of the *S. aureus* bacterium.

It took 10 years from the time when Fleming discovered penicillin before the first clinical trials were successful. These clinical trials proved to everyone that penicillin cured diseases caused by *S. aureus*. The next challenge was to mass-produce penicillin. This required a highly productive strain of *Penicillium*. The breakthrough came with the isolation of a highly productive strain of *Penicillium* isolated from a cantaloupe.

Many of the antibiotics in use today are produced from *Streptomyces* species. *Streptomyces* are bacteria that live in the soil. Other antibiotics come from the genus *Bacillus* and the genera *Cephalosporium* and *Penicillium*, which are molds.

Antimicrobial Activity: Who to Attack?

The way in which an antibiotic attacks a pathogenic microorganism is called the mode of *antimicrobial activity*. You might say this is the way that the antibiotic identifies the "good guys" from the "bad guys." The good guys are eukaryotic cells, and the bad guys are prokaryotic cells (bacteria). Human cells do not resemble bacterial cells chemically. This is the reason why the antibiotic can differentiate between the good guys and the bad guys.

Eukaryotic cells and prokaryotic cells are different in a number of ways, such as the presence or absence of cell walls and their chemical makeup. There are also differences in their respective metabolisms and structures of their organelles, such as the ribosome. It is these differences that antibiotics target so that only prokaryotic cells are destroyed.

Sometimes there can be a problem, especially when the disease-causing agents are other eukaryotic cells. While bacterial cells are dissimilar to human cells, the same cannot be said of other pathogenic microorganisms such as helminths, fungi, and protozoa. These microorganisms are comprised of eukaryotic cells that resemble human cells. Therefore, antibacterial agents are not effective against helminths, fungi, and protozoa. Likewise, antibiotics are of no use against viruses because a virus invades a human cell and reprograms them with the genetic information of the virus to create new viruses. Since the virus is inside the human cell, an antibiotic is ineffective.

Spectrum of Antimicrobial Activity: Some Bad Guys Get Away

The number of different types of pathogenic microorganisms that an antibiotic can destroy is called the *spectrum of antimicrobial activity*. Therefore, antibiotics are referred to as *broad-spectrum* or *narrow-spectrum* agents.

A *broad-spectrum antibiotic* is an antibiotic that destroys many types of bacteria, such as both gram-positive and gram-negative bacteria. A *narrow-spectrum antibiotic* is an antibiotic that destroys a few types of bacteria, such as only gram-negative bacteria.

The deciding factor in the spectrum of antimicrobial activity is porins in the lipopolysaccharide outer layer of gram-negative bacteria. A *porin* is a water-filled channel that forms in the lipopolysaccharide outer layer, enabling substances on the outside of the cell to enter the cell.

In order for an antibacterial drug to destroy a bacterium, the drug must enter the bacterial cell through the porin channel. However, to do so, the drug must be relatively small and *hydrophilic*, which means that the antibacterial drug has an affinity for water (which is contained in the porin channel). Some drugs are relatively large or are lipophilic, which means that the antibiotic has an affinity for lipids and is attracted to the lipopolysaccharide outer layer of the cell (rather than the water in the porin channel).

The Battle of the Pathogens: Some Bad Guys Are Good Guys, Too

Our bodies contain many microorganisms that are normal and beneficial. Others simply are unable to grow to the level where they become pathogenic because they compete with other microorganisms for the nutrients required for growth.

This situation causes scientists concern when giving a broad-spectrum antibiotic to a patient when the pathogen is not known. A *pathogen* is a disease-causing organism. A broad-spectrum antibiotic is likely to destroy the pathogen, but it is also likely to destroy other normal flora. This could cause the pathogen to become opportunistic, making the host extremely ill. An example of this would be a *Clostridium difficile* infection. The increased growth of opportunistic pathogens is called *superinfection*. Microorganisms that develop resistance to the antibiotic also cause a superinfection by replacing the antibiotic-sensitive strain.

The Attack Plan

As discussed earlier in this chapter, an antimicrobial drug uses one of two strategies to combat a pathogen—either a bactericidal strategy or a bacteriostatic strategy. The *bactericidal* strategy is a direct hit, killing the pathogen and preventing it from spreading. The *bacteriostatic* strategy prevents the growth of bacteria. In bacteriostasis, the host's immune system fights the pathogen through phagocytosis and the production of antibodies.

The Bacteriostatic Strategy

One of the first targets of attack of the bacteriostatic strategy is the cell wall of the pathogen. The objective is to weaken the cell wall, causing the cell to undergo lysis. The key to this attack is the structure of the cell wall itself. Bacterial cell walls are comprised of a network of macromolecules called *peptidoglycans*. Certain antibiotics inhibit the making of peptidoglycans, thus weakening the cell wall. This allows water to enter the cell resulting in lysis. Antibiotics that affect the synthesis of the cell wall of bacteria are bacitracin, vancomycin, penicillin, and cephalosporins.

Attacking Protein Synthesis

Another target of attack of the bacteriostatic strategy is the pathogen's ability to make protein. Proteins are necessary for both eukaryotic and prokaryotic cells. If the antibiotic can inhibit protein synthesis, then the cell dies. The problem is for the antibiotic to identify only prokaryotic cells (bacteria) and not eukaryotic cells, which include human cells.

The solution lies within the structure of ribosomes in eukaryotic and pro-karyotic cells. Ribosomes are the site of protein synthesis.

Eukaryotic cells have 80S ribosomes, and prokaryotic cells have 70S ribo-somes. The numbers are referred to as Svedberg units and indicate the relative sedimentation rates during centrifugation. Prokaryotic ribosomes consist of a small subunit referred to as a *30S subunit* and a large subunit called a *50S sub-unit*. The 30S subunit contains one molecule of ribosomal RNA (rRNA), and the 50S subunit contains two molecules of rRNA.

Antibiotics use the differences in ribosomes to distinguish prokaryotic cells from eukaryotic cells, thereby interfering with protein synthesis only for the prokaryotic cells. Some antibiotics interfere with the 50S subunit, whereas other antibiotics attack the 30S subunit. These antibiotics include chloram-phenicol, erythromycin, streptomycin, and tetracycline.

Chloramphenicol interferes with the 50S subunit by preventing peptide bonds from forming. *Erythromycin* is a narrow-spectrum antibiotic that also interferes with the 50S subunit but only for gram-positive bacteria. *Tetracy-cline* interferes with the 30S subunit and prevents the transfer RNA (tRNA) from carrying amino acids and prevents amino acids from attaching to the polypeptide chain. *Streptomycin* interferes with the 30S subunit by changing its shape, causing an incorrect reading of the genetic code on the messenger RNA (mRNA). Streptomycin is an example of what is called an *aminoglycoside* anti-biotic. (An aminoglycoside is made up of amino carbohydrates and an amino-cyclitol ring.)

Attacking the Plasma Membrane

Still another target of attack by antibiotics is the pathogen's *plasma membrane*. The plasma membrane is permeable, allowing substances in and out of the cell as part of normal cell metabolism.

Some antibiotics change the permeability of the plasma membrane, thereby disrupting the metabolism of the pathogen. One such antibiotic is *polymyxin B*, which attaches to the phospholipids of the plasma membrane, inhibiting per-meability of the membrane.

Antifungal drugs also destroy fungi using a similar technique. The plasma membranes of fungi are made up predominately of *ergosterol*. Antifungal drugs combine with the sterols to inhibit the permeability of the membrane. Popular antifungal drugs include *amphotericin B*, *ketoconazole*, and *miconazole*.

Attacking Synthesis of Nucleic Acid

Nucleic acids are blueprints for the reproduction of every cell; these are DNA and RNA. A commonly used bacteriostatic strategy is to interfere with the making of nucleic acid by using *rifampin*, *quinolones*, and other similar antibiotics.

Although disrupting the formation of nucleic acid results in destruction of the pathogen, scientists are careful when choosing the antibiotic for this purpose because the antibiotic might interfere with the host's DNA and RNA.

Attacking Metabolites

A *metabolite* is a substance (such as an enzyme) that is necessary for cell metabolism. The bacteriostatic strategy interferes with metabolites and prevents the growth of the pathogenic organism.

For example, some pathogens require the substrate *para*-aminobenzoic acid (PABA) in order to synthesize folic acid. Folic acid is a coenzyme that is involved in the synthesis of purine and pyrimidine nucleic acid bases and many amino acids.

The antimetabolite *sulfanilamide*, which is a sulfa drug, resembles PABA. When sulfanilamide is introduced into the pathogenic microorganism, the enzyme used in making folic acid combines with sulfanilamide instead of PABA. This then disrupts the formation of folic acid and eventually prevents the pathogenic microorganism from synthesizing purine and pyrimidine.

Exploring Antimicrobial Drugs

Antimicrobial drugs are classified by their antimicrobial activity. These classifications are cell wall inhibitors, protein inhibitors, plasma membrane inhibitors, nucleic acid inhibitors, antimetabolites, antifungal drugs, antiviral drugs, enzyme inhibitors, antiprotozoan drugs, and antihelminthic drugs. Let's take a close look at each of these classifications.

Cell Wall Inhibitors

A *cell wall inhibitor* is an antimicrobial drug that inhibits the growth or functionality of the cell wall of a pathogen. The more common cell wall inhibitors follow.

Penicillin

Penicillin is a group of antibiotics that have a *β-lactam ring* in the core structure. Each member of the group has a different side chain that is attached to the β-lactam ring. Penicillin can be natural or semisynthetic.

- *Natural penicillin.* Natural penicillin is extracted from the *Penicillium* mold. The major disadvantage of natural penicillin, except for Penicillin V, is that it is negatively affected by stomach acid, meaning that the most effective way to administer natural penicillin is via an intramuscular injection. Another problem is that penicillin itself can be attacked by the enzyme penicillinase. *Penicillinases,* also known as *β-lactamases,* are enzymes produced by many bacteria that attach to the β-lactam ring of the penicillin, rendering it ineffective. The more common natural penicillins include the following:
  - *Penicillin G* is the prototype compound for natural penicillin, as is used in defense of staphylococci, streptococci, and several spirochetes. Natural penicillin has a narrow spectrum of activity. Penicillin G lasts up to 6 hours when injected intramuscularly.
  - *Procaine penicillin* is a combination of the drugs procaine and penicillin G and lasts up to 24 hours, although the concentration of procaine penicillin diminishes after 4 hours.
  - *Benzathine penicillin* is a combination of benzathine and penicillin G and can last for up to 4 months.
  - *Penicillin V* is not inhibited by acid in the stomach and can be taken orally.
- *Semisynthetic penicillin.* Semisynthetic penicillins are chemically altered natural penicillins that are designed to overcome disadvantages of natural penicillins. There are two ways in which scientists modify natural penicillins. They stop the natural synthesis of a *Penicillium* molecule and use the penicillin core structure. The other way is to modify the side chains from the natural penicillin and insert a new side chain that overcomes the disadvantage of the original side chain. This is the technique used to combat penicillinase. Common types of semisynthetic penicillin include the following:
  - *Methicillin* was the first semisynthetic penicillin, developed to resist penicillinase.
  - *Oxacillin* is a newer semisynthetic penicillin that resists penicillinase and has replaced methicillin.
  - *Ampicillin* is a semisynthetic penicillin that overcomes the narrow spectrum of activity of natural penicillin by attacking gram-negative and gram-positive bacteria. Ampicillin is not resistant to penicillinases.

- ◦ *Amoxicillin* is similar to ampicillin.
- ◦ *Carbenicillin* is a member of the carboxypenicillin group that has a broad spectrum of activity against gram-negative bacteria and is used to fight *Pseudomonas aeruginosa*.
- ◦ *Ticarcillin* is similar to carbenicillin.
- ◦ *Mezlocillin* is a member of the ureidopenicillins group of penicillins and is a modification of ampicillin, giving it a broader spectrum of activity against bacteria.
- ◦ *Azlocillin* is similar to mezlocillin.
- ◦ *Augmentin* (which is the trade name) is a combination of amoxicillin and potassium clavulanate, which is produced by *Streptomyces*. Amoxicillin attacks gram-negative and gram-positive bacteria, and potassium clavulanate resists penicillinase.

Still Struggling

Still having a problem remembering penicillin drugs? This summery can help: *Methicillin* was the first semisynthetic penicillin. *Oxacillin* is a newer semisynthetic penicillin that resists penicillinase and has replaced methicillin. *Ampicillin* is a semisynthetic penicillin that attacks gram-negative and gram-positive bacteria. Ampicillin is not resistant to penicillinases. *Amoxicillin* is similar to ampicillin. *Carbenicillin* is a member of the carboxypenicillin group that works against gram-negative bacteria and is used to fight *P. aeruginosa*. *Ticarcillin* is similar to carbenicillin. *Mezlocillin* is a member of the ureidopenicillins group of penicillins and is a modification of ampicillin, giving it a broader spectrum of activity against bacteria. *Azlocillin* is similar to mezlocillin. *Augmentin* (which is the trade name) attacks gram-negative and gram-positive bacteria.

Monobactams

Monobactams are synthetic antibiotics that have a single lactam ring and resist penicillinase. Monobactams destroy only aerobic gram-negative bacteria such as *Escherichia coli*.

Cephalosporins

Cephalosporins are resistant to penicillinases and destroy gram-negative bacteria. The disadvantage of cephalosporins is they can be made ineffective by β-lactamases.

Carbapenems

Carbapenems have a broad spectrum of activity that prevents breakdown of the antibiotic by the kidneys. These antibiotics inhibit the synthesis of cell walls. An example is Primax (trade name). Primax has proven effective against 98 percent of all organisms that were isolated from patients in hospitals.

Bacitracin

Bacitracin is used against staphylococci and streptococci and other gram-positive bacteria and is used as a topical antibiotic for superficial infections.

Vancomycin

Vancomycin has a very narrow spectrum of activity and is the most effective antibiotic against staphylococci that produce penicillinase. Vancomycin is used in the treatment of endocarditis. However, vancomycin can be toxic to humans.

Isoniazid

Isoniazid (INH) inhibits mycolic acid, which is needed for synthesizing the cell wall of *Mycobacteria tuberculosis*. Isoniazid is only effective in fighting mycobacterial tuberculosis.

Ethambutol

Ethambutol is similar to isoniazid and is used as a secondary treatment to avoid *M. tuberculosis* becoming resistant to isoniazid.

Protein Inhibitors

A *protein inhibitor* interferes with a pathogen's ability to synthesize protein. The commonly used protein inhibitor drugs follow.

Aminoglycosides

Aminoglycosides are a group of antibiotics that were the first to attack gram-negative bacteria. Their disadvantage is that they can cause permanent damage to the auditory nerve and cause kidney damage. Commonly used members of this group of antibiotics include the following:

- *Streptomycin* is used as a secondary treatment for tuberculosis. It was discovered in 1944 and today is used sparingly because of its toxic effect on humans and the fact that bacteria quickly become resistant to it.
- *Neomycin* is a topical antibiotic used for superficial infections.
- *Gentamicin* is an antibiotic used to combat *P. aeruginosa* infections.

Tetracyclines

Tetracycline is a broad-spectrum antibiotic that attacks both gram-positive and gram-negative bacteria. Tetracyclines are able to combat pathogens that invade cells because they can enter body tissues. They are used against urinary tract infections, rickettsial infections, and chlamydial infections. Tetracyclines are also used as the secondary treatment for gonorrhea and syphilis. The disadvantages of tetracyclines are discoloration of teeth in children and liver damage in pregnant women. Common members of the tetracycline group of antibiotics include the following:

- *Oxytetracycline* is a commonly used form of tetracycline that is also known as *Terramycin*.
- *Chlortetracycline* is another commonly used form of tetracycline that is also known as *Aureomycin*.
- *Doxycycline* is a semisynthetic form of tetracycline that has a longer-lasting effect than the natural form of tetracycline.
- *Minocycline* is a semisynthetic form of tetracycline.

Chloramphenicol

Chloramphenicol is a broad-spectrum antibiotic that is small enough to affect areas of the body that are too small for other antibiotics to enter. Chloramphenicol is an antibiotic of last resort because it inhibits the formation of blood cells, causing aplastic anemia.

Macrolides

Macrolides are a group of antibiotics that target the cell wall. They affect the organism by inhibiting its ability to synthesize protein; they are a secondary drug when penicillin G is unavailable or the organism has become resistant to penicillin G. Macrolide antibiotics can be administered orally, making them the choice for treating children who have streptococcal and staphylococcal infections. The most commonly used macrolide is *erythromycin*, which is used to treat legionellosis, mycoplasmal pneumonia, and streptococcal and staphylococcal infections.

Plasma Membrane Inhibitors

Plasma membrane inhibitors are antibiotics that interfere with the functionality of the plasma membrane of a pathogen. The commonly used plasma membrane inhibitor antibiotic is *polymyxin B*, which combats gram-negative bacteria such as *Pseudomonas*. Today, when combined with neomycin, it is used in nonprescription topical antibiotics for superficial infections.

Nucleic Acid Inhibitors

Nucleic acid inhibitors interfere with the formation of nucleic acids. The commonly used nucleic acid inhibitor antibiotics follow.

Rifamycins

Rifamycins comprise a group of antibiotics that inhibit mRNA synthesis. They are used to treat tuberculosis and leprosy. Rifamycins can reach the cerebrospinal fluid and enter tissues and abscesses. The disadvantage of rifamycins is that they cause urine, feces, saliva, sweat, and tears to appear an orange-red color. A commonly used rifamycin is *rifampin*, which is used to attack mycobacterium that cause tuberculosis and leprosy.

Quinolones

Quinolones interfere with DNA gyrase (an enzyme), which is involved in DNA replication. Quinolones are used only for infections of the urinary tract.

Fluoroquinolones

Fluoroquinolones comprise a group of antibiotics that have a broad spectrum of activity and target urinary tract infections. Fluoroquinolones are able to enter cells to attack pathogens. The disadvantage of fluoroquinolones is that they may interfere with cartilage development, making them unsafe for pregnant women and children. Two examples that are used most commonly are *norfloxacin* and *ciprofloxacin*.

Antimetabolites

Antimetabolites are a group of drugs that interfere with metabolites, which are necessary for various reactions in the cell. Common antimetabolites follow.

Sulfonamides

Sulfonamides, or *sulfa drugs*, are used to treat urinary tract infections and control infections in burn patients. These were among the first synthetic antimicrobial drugs used. Commonly used sulfonamides include the following:

- *Silver sulfadiazine* is used to control infections common in burn patients.
- *Trimethoprim-sulfamethoxazole* (TMP-SMZ) is a combination of trimethoprim and sulfamethoxazole that is used to combat intestinal and urinary tract gram-negative pathogens.

Antifungal Drugs

Antifungal drugs inhibit the growth of fungi. The commonly used antifungal drugs follow.

Polyenes

Polyenes are a group of antifungal antibiotics that combine with the ergosterol in the plasma membranes of fungi. This causes the plasma membrane to become extremely permeable, ultimately resulting in death of the fungus. A commonly used polyene is *amphotericin B*, which is used to treat histoplasmosis, coccidioidomycosis, and blastomycosis. However, amphotericin B is toxic to the kidneys. Enclosing amphotericin B in liposomes when administering the drug reduces its toxicity.

Imidazoles

Imidazoles interfere with fungal ergosterol synthesis. Commonly used imidazoles include the following:

- *Clotrimazole* is used to treat cutaneous mycoses. Two examples are athlete's foot and vaginal yeast infections.
- *Miconazole* is similar to clotrimazole. Both drugs are used topically and sold without a prescription.

Triazoles

Triazoles are similar to imidazoles but are less toxic than other antifungal agents.

Griseofulvin

Griseofulvin is an antimicrobial used to treat dermatophytic fungal infections of the hair and nails, including tinea capitis. Griseofulvin is taken orally and enters the keratin of skin, hair, and nails. Its purpose is to interfere with fungal mitosis, inhibiting reproduction.

Tolnaftate

Tolnaftate is a topical treatment for athlete's foot. Its mechanism of action is unknown.

Antiviral Drugs

Antiviral drugs interfere with the replication of viruses. Commonly used antiviral drugs follow.

Amantadine

Amantadine prevents a virus from entering a host cell or from uncoating once the virus enters the cell. It is used (although with limited usefulness) to prevent influenza A but has no practical effect once the virus infects the cell.

Nucleoside Analogues

Nucleoside analogue antiviral drugs affect the synthesis of viral DNA or RNA. Commonly used nucleoside analogues include the following:

- *Acyclovir* is used to combat the viruses that cause herpes.
- *Ribavirin* is used to treat rotavirus-caused pneumonia in infants.
- *Ganciclovir* is used to fight cytomegalovirus infections that are common in transplant patients and patients who have AIDS.
- *Trifluridine* is used to treat the eye infection caused by acyclovir-resistant herpes keratitis.
- *Zidovudine* (AZT) is used in HIV infection. This drug blocks the synthesis of DNA from RNA by the enzyme reverse transcriptase.

Antiprotozoan Drugs

Antiprotozoan drugs are used to combat parasitic infections such as the disease malaria. Commonly used antiprotozoan drugs follow.

Chloroquine

Chloroquine is a replacement for quinine, which was the traditional treatment for malaria.

Mefloquine

Mefloquine is a secondary treatment for malaria when there is a resistance to chloroquine.

Quinacrine

Quinacrine is used to treat giardiasis.

Diiodohydroxyquin

Diiodohydroxyquin is used to treat intestinal amoebic diseases. However, the drug can cause damage to the optic nerve if the dosage is not carefully controlled.

Metronidazole

Metronidazole combats parasitic protozoa and obligate anaerobic bacteria. It is used to treat vaginitis caused by *Trichomonas vaginalis*, giardiasis, and amoebic dysentery.

Nifurtimox

Nifurtimox is used to combat Chagas disease. However, nifurtimox can cause side effects, such as nausea and convulsions.

Antihelminthic Drugs

Antihelminthic drugs are used to combat helminths, which are parasitic flat-worms. Commonly used antihelminthic drugs follow.

Niclosamide

Niclosamide is used to treat tapeworm infections. This drug inhibits adenosine triphosphate (ATP) production in aerobic conditions.

Praziquantel

Praziquantel is also used to destroy tapeworm infections and is used to treat schistosomiasis and other fluke-caused diseases. This drug alters the permeability of the organism's plasma membrane.

Mebendazole

Mebendazole is used to combat intestinal helminthic infections such as ascariasis, whipworms, and pinworms.

Chemotherapy Tests (Antibiotic Sensitivity Tests)

Chemotherapeutic Tests are scientific tests that determine which antibiotic combats a particular pathogen. These tests are used to determine which antibiotic to prescribe to treat a specific disease. The antibiotic is called a *chemotherapeutic agent*.

There are two popular chemotherapy tests: diffusion methods and the broth-dilution method.

Diffusion Methods

Diffusion methods are tests that place a paper disk or plastic strip that is coated with a chemotherapeutic agent in touch with a pathogen to determine if the agent inhibits the pathogen. There are two commonly used diffusion methods. These are the disk-diffusion method and the minimal inhibitory concentration method.

The *disk-diffusion method*, also known as the *Kirby-Bauer test*, uses a filter paper disk that is coated with the chemotherapeutic agent being tested. The paper disk is placed in a Petri dish that contains an agar medium that has been contaminated with the pathogen. The Petri dish then is incubated, during which time the chemotherapeutic agent diffuses into the agar. A zone of inhibition can be seen in the agar medium. A *zone of inhibition* is an area where the pathogen does not grow because the chemotherapeutic agent inhibits it. Scientists measure the diameter of the zone of inhibition and compare the diameter with a standard table for the chemotherapeutic agent. The table indicates whether the pathogen is sensitive, intermediately sensitive, or resistant to the chemotherapeutic agent.

The *minimal inhibitory concentration (MIC) test* is an advanced diffusion method that determines the lowest concentration of chemotherapeutic agent that inhibits visible growth of bacteria. This test uses a plastic-coated strip that contains a gradient concentration of the chemotherapeutic agent. The strip then is exposed to the pathogen, after which the strip is compared with a printed scale to determine the minimal inhibitory concentration of the chemotherapeutic agent.

Broth-Dilution Method

The *broth-dilution method* is used to determine the minimal inhibitory concentration and the minimal bactericidal concentration (MBC) of the chemotherapeutic agent. The *minimal bactericidal concentration* is the lowest concentration of a chemotherapeutic agent needed to kill the pathogen.

The broth-dilution test requires that a broth containing the drug be placed in wells of a plastic tray in a sequence of decreasing concentrations of drug. Each well is inoculated with the bacteria. After an incubation period, each well is examined to determine the effectiveness of the concentration of the chemotherapeutic agent. The well that shows no pathogen growth with the lowest concentration of the chemotherapeutic agent signifies the MIC and MBC that should be used to treat the disease caused by the pathogen.

Clinical laboratories use an automated broth-dilution test where a computer scans wells and reports results.

QUIZ

Match the antifungal drug:

    A. Amphotericin B
    B. Griseofulvin
    C. Tolnaftate

1. This antifungal drug is used to treat histoplasmosis and coccidioidomycosis.
_____

2. This antifungal drug is used to treat athlete's foot. _____

3. This antifungal drug is used to treat dermatophytic fungal infections. _____

Match the antiprotozoan drug:

    A. Chloroquine
    B. Mefloquine
    C. Quinacrine
    D. Diiodohydroxyquin
    E. Metronidazole

4. This antiprotozoan drug is used to treat giardiasis. _____ or _____

5. This antiprotozoan drug is used to treat vaginitis caused by *Trichomonas vaginalis.* _____

6. This antiprotozoan drug is used to treat intestinal amoebic diseases. _____

7. This antiprotozoan drug is used to treat malaria. _____

8. This antiprotozoan drug is used as a secondary treatment for malaria. _____

Match the antihelminthic drug:

    A. Niclosamide
    B. Mebendazole

9. This antihelminthic drug is used to treat intestinal helminthic infections such as whipworms and pinworms. _____

10. This antihelminthic drug is used to treat tapeworm infections. _____

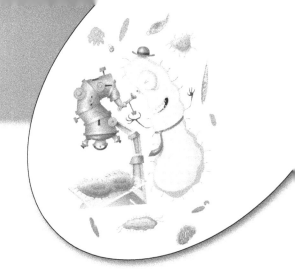

Final Exam

Match the organism:

    A. Hansen's disease

    B. Impetigo

    C. Gastric ulcers

    D. Black plague

    E. Gas gangrene

    F. Lock jaw

    G. African sleeping sickness

    H. Meningitis

    I. Botulism/Botox

    J. Giardiasis

1. *Trypanosoma gambiense* _____

2. *Giardia lamblia* _____

3. *Mycobacterium leprae* _____

4. *Clostridium botulinum* _____

5. *Clostridium tetani* _____

6. *Clostridium perfringens* _____

7. *Staphylococcus aureus* _____

8. *Yersinia pestis* _____

9. *Neisseria meningitides* _____

10. *Helicobacter pylori* _____

Fill in the blanks:

11. An organism that can cause a disease is called a _____.

12. Microorganisms that can grow with or without oxygen are called _____.

13. A sterile culture medium contains _____.

14. A solidifying agent used in cultures is called _____.

15. This culture provides growth of a certain microorganism but not others _____.

16. By hydration and use of liquid nutrient media, bacterial cultures can be _____.

17. Organisms that possess a nucleus, nuclear envelope, cytoplasm, and organelles are called _____.

18. The "energy currency" molecule of an organism _____.

19. Small molecules that combine to form large molecules use this type of reaction. _____.

20. Reactions that give off energy as molecules are broken down are called _____.

21. The energy required to begin a chemical reaction is called _____.

22. A catabolic reaction that breaks down glucose is called _____.

23. Very high or very low pH will cause the _____ of enzymes.

24. The spontaneous decaying and giving off of particles of radium is called _____.

25. The outer most shell of an atom is called _____.

26. What kind of reaction breaks the bond between atoms of a molecule or compound? _____

27. Reactions that perform synthesis and decomposition reactions are called _____.

28. An increase in temperature, pressure, and element orientation will increase _____.

29. This membrane provides a selective barrier between the nucleus and the cell's internal structures. _____

30. Movement of molecules through the plasma membrane is regulated by _____.

31. What forms channels in the plasma membrane and permits the flow of molecules through the plasma membrane? _____

32. The movement of water from a region of higher concentration to a region of lower concentration is called _____.

33. When substances move from a region of higher concentration to a region of lower concentration, it is called _____.

34. The movement of a substance across the plasma membrane against the gradient using energy employs this type of transport. _____

35. The condition of a cell "eating" is called _____.

True (T) or false (F):

36. Oxygen is not required for a facultative anaerobe to grow. _____

37. A reaction that breaks the bond between atoms of a molecule or compound is called a synthesis reaction. _____

38. A chemically defined medium must contain growth factors. _____

39. A microorganism is a small organism that takes in and breaks down chemicals for energy, excretes wastes, and is capable of reproduction. _____

40. Pyruvic acid must be converted to acetyl-coenzyme A before entering the Kreb's cycle. _____

41. The smallest particle of an element is called an atom. _____

42. An inoculating loop is used to "streak plate" an organism over a nutrient medium. _____

43. An enzyme temporarily bonds to the substrate and then positions the substrate to increase the likelihood that the substrate will collide with another atom, ion, or molecule and thereby bring about a chemical reaction. _____

44. Phagocytosis is the process by which cells excrete waste products through their plasma membranes. _____

45. Germ theory states that germs can cause disease in other organisms. _____

46. Glucose is the preferred energy-storage molecule. _____

47. A pathogenic microorganism is a microorganism that causes disease in a host. _____

48. Microorganisms that do not have a distinct nucleus are called prokaryotes. _____

49. Bacteria would be an example of a prokaryotic organism. _____

50. The two nucleic acids found in a cell are called deoxyribonucleic acid and ribonucleic acid. _____

51. No energy is required with passive transport of substances across the plasma membrane. _____

52. No energy is required with active transport of substances across the plasma membrane. _____

53. The "bending" of light rays is called refraction. _____

54. The body of a fungus is called its soma. _____

55. Identification is the process of observing and classifying organisms into a standard group. _____

56. A genus consists of one or more lower ranks called species. _____

57. Taxonomy is the classification of organisms based on a scientifically proven natural relationship. _____

58. The Southern blot technique is used for detecting specific restriction fragments. _____

59. An RNA polymerase is an enzyme that is used in the synthesis of RNA. _____

60. The three nucleotide units within an mRNA strand are called exons. _____

Match the following:

    A. Chemoheterotrophs

    B. Sulfate-reducing archaea

    C. Helical

    D. Pleomorphic

    E. Vibroid

    F. *Mycoplasma*

    G. Hemolysis

    H. Aerobic/microaerophilic

    I. *Halobacterium salinarium*

    J. Extreme halophiles

    K. Methogenic archaea

    L. Photoautotrophs

    M. Pseudopods

    N. Chlamydias

    O. Sulfate reducers

    P. Crenarchaeota

    Q. Schizogony

    R. Cell wall–less archaea

S. Proglottis

T. Extremely thermophilic sulfur metabolizers

61. An organism that has various shapes. _____

62. A single-cell archaea that produces methane. _____

63. Archaea that function in the presence of air. _____

64. Archaea that live in an extremely salty environment. _____

65. Archaea that do not have a cell wall. _____

66. Archaea that need sulfur for growth. _____

67. Achaea that live in the absence of oxygen. _____

68. Archaea that absorb nutrients from dead organic matter. _____

69. Archaea that acquire energy through photosynthesis. _____

70. These archaea extract electrons from marine hydrothermal vents. _____

71. These bacteria require small amounts of oxygen to grow. _____

72. These bacteria have a spiral shape. _____

73. These bacteria have a curved shape. _____

74. Intracellular parasites that need a host in order to reproduce. _____

75. Organisms that resemble fungi. _____

76. Destruction of red blood cells. _____

77. Organisms that obtain energy from sunlight. _____

78. Meaning "false feet." _____

79. Meaning "multiple fission." _____

80. Compartments in a tapeworm that contain reproductive organs. _____

Multiple choice:

81. **Describe a gram-positive cell wall.**
    A. A gram-positive cell wall has many layers that repel the crystal of violet dye when the cell is stained.
    B. A gram-positive cell wall has many layers of peptidoglygan that retain the crystal of violet dye when the cell is stained.
    C. A gram-positive cell wall has one layer that retains the crystal of violet dye when the cell is stained.
    D. A gram-positive cell wall has one layer of peptidoglygan that repels the crystal of violet dye when the cell is stained.

82. **A microscope that has two sets of lenses is called a**
    A. compound microscope.
    B. bright-field microscope.
    C. electron microscope.
    D. phase-contrast microscope.

83. **These proteins help to support the plasma membrane.**
    A. Integral proteins
    B. Transmembrane proteins
    C. Peripheral proteins
    D. Lipoproteins

84. **Immunity where antigens are injected into an organism as a vaccine is called**
    A. naturally acquired active immunity.
    B. naturally acquired passive immunity.
    C. artificially acquired active immunity.
    D. artificially acquired passive immunity.

85. **What enables scientists to take nucleotide fragments from other DNA and reassemble fragments into a new nucleotide sequence?**
    A. Recombinant DNA technology
    B. Enzyme technology
    C. Enzyme DNA technology
    D. Recombinant enzyme technology

86. Archaea that function in the presence of air are called
    A. extreme halophiles.
    B. sulfate reducers.
    C. methanogenic archaea.
    D. cell wall–less archaea.

87. What is another name for a restriction endonuclease?
    A. Plasmid
    B. Vector
    C. Agarose gel
    D. Restriction enzyme

88. What acquires nutrients through photosynthesis?
    A. Humans
    B. Animals
    C. Fungi
    D. Plants

89. Which of the following is not a type of RNA?
    A. uRNA
    B. rRNA
    C. tRNA
    D. mRNA

90. Translation begins the synthesis of which of the following?
    A. Carbohydrates
    B. Lipids
    C. Proteins
    D. Ribosomes

91. The way in which an antibiotic attacks a pathogen is called
    A. lipophilic.
    B. hydrophilic.
    C. antimicrobial activity.
    D. broad spectrum.

92. A small infectious particle that is composed of a protein is called a

    A. prion.

    B. capsid.

    C. viron.

    D. viroid.

93. What is a capsid?

    A. A capsid is the membrane bilayer of a virus.

    B. A capsid is the protein coat that encapsulates a virus.

    C. A capsid is another name for a bacteriophage.

    D. A capsid is the envelope around a virus.

94. A free virus particle is

    A. another name for a virus.

    B. a synthesized virus.

    C. a viron.

    D. a mature virus.

95. Pieces of the host cell's membrane make

    A. the capsid.

    B. the nucleus of the host cell.

    C. the nucleus of the virus.

    D. the envelope of a virus.

96. What type of virus is the human papillomavirus?

    A. Plant

    B. Oncogenic

    C. Viroid

    D. Prion

97. Some protists use these structures to move about in their environment.

    A. Mitochondria

    B. Mesoderm

    C. Vacuole

    D. Cilia

98. A platyhelminth is a

    A. flat worm.

    B. earth worm.

    C. nematode.

    D. roundworm.

99. Unicellular algae that have the capacity of self-movement by a flagellum are called

    A. dinoflagellates.

    B. diatomes.

    C. chrysophytes.

    D. lichens.

100. These unicellular algae have a hard shell.

    A. Chrysophytes

    B. Dinoflagellates

    C. Phaeophyta

    D. Diatomes

Answers to Quizzes and Final Exam

Chapter 1
1. agar
2. scientific method
3. hypothesis
4. experiment
5. theory
6. B
7. A
8. C
9. E
10. D

Chapter 2
1. nucleus
2. electrons
3. protons
4. orbit
5. atomic number
6. B
7. D
8. E
9. A
10. C

Chapter 3
1. Hans Christian Gram
2. Ziehl and Neelsen
3. endospore stain
4. refractive index
5. resolution
6. E
7. A
8. D
9. B
10. C

Chapter 4
1. responsiveness
2. reproduction
3. movement
4. growth
5. metabolism
6. D
7. A
8. E
9. B
10. C

Chapter 5
1. nicotinamide adenine dinucleotide
2. flavine adenine dinucleotide
3. nicotinamide adenine dinucleotide phosphate
4. adenosine triphosphate
5. adenosine diphosphate
6. B
7. D
8. E
9. C
10. A

Chapter 6
1. sterilization
2. antisepsis

3. commercial sterilization
4. disinfection
5. degerming
6. E
7. A
8. B
9. D
10. C

Chapter 7
1. mutation rate
2. Francois Jacob and Jacques Monod
3. transcription
4. translation
5. G
6. A
7. F
8. B
9. C
10. D
11. E

Chapter 8
1. genetic engineering
2. genetic material
3. restriction fragment length polymorphosis
4. recombinant DNA technology
5. recombinant DNA
6. B
7. D
8. A
9. E
10. C

Chapter 9
1. Carl Linnaeus
2. five kingdoms
3. Monera
4. Plantae
5. Eukarya
6. D
7. B
8. C and E
9. F
10. A

Chapter 10
1. *Yersinia pestis*
2. pneumonic plague
3. *Vibrio cholerae*
4. *Vibrio parahaemolyticus*
5. *Pasteurella multocida*
6. C
7. A
8. B
9. D
10. E

Chapter 11
1. *Rhizopus nigricans*
2. *Histoplasma*
3. Chrysophytes
4. Dinoflagellates
5. Gonyaulax
6. C
7. B
8. A
9. D
10. E

Chapter 12
1. DNA
2. DNA
3. RNA
4. RNA
5. DNA
6. C
7. B
8. E
9. D
10. A

Chapter 13
1. subclinical carriers
2. chronic carriers
3. convalescent carriers
4. incubatory carriers
5. F
6. B
7. C
8. E
9. D
10. A

Chapter 14
1. Interleukin-8
2. Interleukin-1
3. Interleukin-2
4. Interleukin-12
5. D
6. A
7. C
8. B
9. E
10. F

Chapter 15
1. B
2. G
3. H
4. C
5. I
6. J
7. E
8. F
9. A
10. D

Chapter 16
1. A
2. C
3. B
4. C or E
5. E
6. D
7. A
8. B
9. B
10. A

Final Exam
1. G
2. J
3. A
4. I
5. F
6. E
7. B
8. D
9. H
10. C
11. pathogenic organism
12. facultative anaerobes
13. no living organisms
14. agar
15. selective media
16. revived
17. eukaryotic organisms
18. adenosine triphosphate
19. anabolic reactions
20. catabolic reactions
21. activation energy
22. glycolysis
23. denaturing
24. radioactivity
25. valence shell
26. decomposition reaction
27. exchange reaction
28. reaction rate
29. nuclear membrane
30. membrane proteins
31. integral protein
32. osmosis
33. simple diffusion
34. active transport
35. phagocytosis
36. T
37. F
38. T
39. T
40. T
41. T
42. T
43. T
44. F
45. T
46. T
47. T
48. T
49. T
50. T
51. T
52. F
53. T
54. T
55. T
56. T
57. F
58. T
59. T
60. F
61. D
62. K
63. O
64. J
65. R
66. T
67. P
68. A
69. I
70. B
71. H
72. C
73. E
74. N
75. F
76. G
77. L
78. M
79. Q

80. S

81. D

82. A

83. C

84. C

85. A

86. B

87. D

88. D

89. A

90. C

91. C

92. A

93. B

94. C

95. D

96. B

97. D

98. A

99. A

100. D

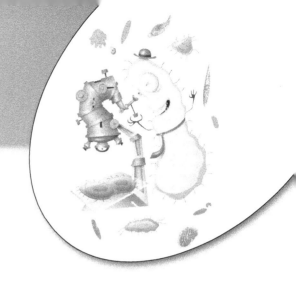

Index

A

Acanthamoeba polyphaga, 183
Acetobacter, 158
Acetyl-CoA, 96
Acid, 35
Acid-base balance, 35
Acid-fast microorganisms, 60
Acid stain, 59
Acquired immunity, 223, 225
Actin, 40
Actinomyces israelii, 71, 168
Actinomycetes, 109, 168
Activation energy, 93
Activation of complement, 230
Active enzyme, 92
Active immunity, 224
Active site, 93
Active transport, 79
Acyclovir, 256
Adenosine diphosphate (ADP), 42
Adenosine triphosphate (ATP), 41–42, 79, 86, 90
ADP. *See* Adenosine diphosphate
Aerial hyphae, 173
Aerobic/microaerophilic, motile, helical/vibroid, gram-negative bacteria, 155
Agar, 14–15, 106, 107
Agarose gel electrophoresis, 133
Agglutination, 229
Agglutination reaction tests, 239–240
Agrobacterium, 157
Airborne pathogen, 212
Alcohols, 115

Aldehydes, 116
Algae, 5, 175–181
    lichens, 181
    reproduction, 176
    types, 176–180
Alpha radiation, 27
Amantadine, 256
Aminoglycoside, 248
Aminoglycosides, 252–253
Ammonia, 93
Amoeba, 182–183
Amoeboid motion, 84
Amoxicillin, 251
Amphotericin B, 255
Ampicillin, 250
Anabolic reaction, 32, 90
Anabolism, 32
Anaerobic gram-negative cocci and rods, 161
Anaerobic growth, 107
Analytical epidemiology, 216
Anamnestic response, 230
Ancyclostoma duodenale, 190
Ancylostoma braziliense, 190
Ancylostoma caninum, 190
Angle of refraction, 50–51
Anion, 29, 35
Anoxygenic photosynthesis, 100
Answers to quizzes/final exam, 271–274
Anthrax, 13, 211
Anthropod, 213
Antibiosis, 245
Antibiotic, 8, 17, 150
Antibiotic sensitivity tests, 258–259

DeMYSTiFieD®

Hard stuff made easy

The DeMYSTiFieD series helps students master complex and difficult subjects. Each book is filled with chapter quizzes, final exams, and user friendly content. Whether you want to master Spanish or get an A in Chemistry, DeMYSTiFieD will untangle confusing subjects, and make the hard stuff understandable.

CPSIA information can be obtained
at www.ICGtesting.com
Printed in the USA
FFHW021653261218
49996959-54702FF

9 780071 761093